ALTERNATIVE
ENERGY

An Introduction to Alternative & Renewable Energy Sources

Mark E. Hazen

ALTERNATIVE ENERGY

An Introduction to Alternative & Renewable Energy Sources

Mark E. Hazen

PROMPT®

PUBLICATIONS

A Division of
Howard W. Sams & Company
Indianapolis, IN
A Bell Atlantic Company

FIRST EDITION, 1996

PROMPT® Publications is an imprint of Howard W. Sams & Company, a Bell Atlantic Company, 2647 Waterfront Parkway, E. Dr., Suite 100, Indianapolis, IN 46214-2041.

International Standard Book Number: 0-7906-1079-5

Library of Congress Catalog Card Number: 96-70754

Acquisitions Editor: Candace M. Hall
Editor: Natalie F. Harris
Assistant Editors: Pat Brady, Karen Mittelstadt
Illustrator: Mark Hazen
Typesetter: Natalie Harris
Layout: Christy Pierce
Cover Design: Phil Velikan

Additional Materials and Acknowledgments:
ACUSON Corporation; Advanced Wind Turbines, Inc.; American Wind Energy Association; Anthony Fiore Construction, Inc.; Do It Homestead; National Renewable Energy Laboratory; Natural Energy Laboratory of Hawaii; Northern Power Systems; Nuclear Energy Institute; Sandia National Laboratories; Sea Solar Power, Inc.; Siemens Solar Industries; Solarex; Southwest Windpower; U.S. Department of Energy; World Power Technologies. Any additional companies, corporations, or individuals who have provided materials for this book are credited within the text.

TABLE OF CONTENTS

PREFACE

This book is not only timely but of great importance to all. It deals with a topic that once was on everyone's mind and was found ubiquitously in the media from newspapers to books, from radio to television. Yes, alternative energy was the hot topic of the early 1970s when oil prices were high and supply was short. It was a time when the U.S. government was offering tax credits on all forms of alternative energy whether applied to the home or a region through a utility company. Then, in the 1980s, many government incentives and funding dried up as a seeming oil abundance returned along with lower prices. A false sense of security set in that resulted in a dangerous delay to the development and needed maturing of the alternative energy industry. Now, once again, we are faced with an ominous reality. The world's energy needs are greater than ever before, as populations increase and Asian countries awaken industrially and economically. The fact is, as the demand for oil-based energy is increasing, the world supply is decreasing.

As we approach the year 2000, we will all become painfully aware of how late it really is. We will see all too vividly how we wasted the time we had while oil was plentiful and cheap. We will see how we misused the temporary security that 50 years (1950 to 2000) of gluttonous oil consumption bought us. We must educate ourselves to the realities of our plight and use our knowledge to influence politicians to renew our resolve toward the development of a strong alternative and renewable energy infrastructure. As you will see in this book, a small foundation has been laid, but we must expand and build upon it.

One of the vanguard organizations in the United States that has been making progress toward renewable energy utilization is the American Wind Energy Association (AWEA). You will learn much about them in this book, so I will save the details for later. I was greatly impressed with a recent article in their *Wind Energy Weekly* newsletter published over the Internet. So much so that I asked them for permission to reprint the article in full for you to read. I believe you too will be impressed:

NEW OIL PRICE SHOCK SEEN LOOMING AS EARLY AS 2000

(Wind Energy Weekly, #684, May 15, 1996, American Wind Energy Association, Washington, DC http://www.econet.org/awea)

If present economic and oil industry trends continue, future price shocks appear likely as early as the year 2000, with the world facing permanent increases in the price of oil, two new studies have concluded.

The first study, THE WORLD OIL SUPPLY 1990-2030, which was completed in late 1995 by the prestigious Geneva, Switzerland-based group Petroconsultants, deals with the realities of the statistics, pointing out that the world is finding only about seven billion barrels of oil each year in a falling trend, while producing 23 billion barrels a year in answer to rising demand. The study describes this situation as a recipe for bankruptcy.

The second report, prepared by Oak Ridge National Laboratory for the Office of Transportation Technology of the U.S. Department of Energy and made public in mid-January, 1996, suggests that the OPEC (Organization of Petroleum Exporting Countries) nations, in control of two-thirds of the world's reserves, will soon have the ability to regain monopoly power in world oil markets.

"Price shocks can be very profitable to oil producers, and consuming nations appear to have developed no adequate defense against them," the report warns.

A summary of the studies' findings was released January 30, 1996, by Fuels for the Future, a Washington, DC-based public relations organization.

Another warning signal came February 5, 1996, as Japan's Ministry of International Trade and Industry (MITI) reported that the level of Japan's crude oil imports from the Middle East reached a new high in 1995 of 78.6 percent of the country's oil imports. Japan imported slightly more than 1.3 billion barrels of oil during the year from the Middle East. Imports from Indonesia and China shrank, partly because of increased consumption within China itself, the Ministry said.

The Petroconsultants study concludes that while the world may not be running out of oil, it is running out of cheap oil. And nothing, it states, contributes more to the high cost of living than high energy prices.

The study emphasizes that a country's or region's peak oil production comes at the midpoint of depletion, when half of its oil has been produced. Then oil production declines to zero at its depletion rate. North American production peaked in 1974, and the world will hit its midpoint around 2000, the consultancy estimates. Although detailed numbers from the Petroconsultants study were not made public, the group previously predicted in 1994 that production would peak in the year 1999 at 65.6 million barrels a day, and decline to 52.6 million barrels a day by 2010. By 2050, the 1994 report said, world oil production would drop to 17.5 million barrels a day, or slightly more than it was in the 1950s.

Interestingly, the Petroconsultants analysis appears to track fairly closely the projections made in the 1950s and 1960s by Dr. M. King Hubbert, the petroleum geologist who first predicted the eventual exhaustion of U.S. and world oil supplies. Hubbert forecast that U.S. production would begin to decline in about 1970 and that world production would crest in 1995. U.S. crude oil output did in fact peak in 1970 at slightly over nine million barrels a day and has declined substantially since.

Dr. Colin Campbell, another oil analyst and author of THE GOLDEN CENTURY OF OIL, 1950-2050: THE DEPLETION OF A RESOURCE, pegs cumulative world oil production through 1993 at 718 billion barrels, with remaining reserves at 932 billion barrels. At the world's current production rate of 23 billion barrels a year, the midpoint between past output and remaining reserves (about 826 billion barrels) should be crossed sometime before the end of the century.

The problem of gradually tightening world oil supplies is exacerbated by a growing concentration of remaining reserves in the Persian Gulf. All other producing countries but the five Persian Gulf states (Saudi Arabia, Iran, Iraq, Kuwait, and the United Arab Emirates) will peak before 2000. The Oak Ridge report points out that by the turn of the century only OPEC will have the capability of developing and producing energy in the quantity required. The report says the problem is not one of the United States running out of oil, but that a handful of nations, which today control about two-thirds of the world's reserves, will have the monopolistic power of a cartel as early as the turn of the century.

During the 1970s, the OPEC cartel was able to raise prices dramatically, extracting billions of dollars from consuming nations. However, higher prices spurred drilling activity worldwide, and newer technologies reduced consumption to the point that the world had excess capacity for more than a decade. Now, supply and demand are nearly in balance, and the advantage is swinging toward the oil producers. With world demand increasing and the reserves of most producers gradually diminishing, the OPEC cartel with its many resources appears ready to control the market for oil during the early days of the new century, said the Oak Ridge analysts.

The Oak Ridge report, in emphasizing the problem, points out that while farm commodities can increase production within a year, it takes 10 to 20 years to develop and produce oil, lending far greater power to an oil cartel's monopoly in the short run.

The Petroconsultants study suggests that few new petroleum sources will be found in the future, and that therefore, most oil production will have to come from existing fields. Estab-

lished oil well reserves are currently at about one trillion barrels, but the report does not make the error of dividing this figure by current production to suggest wrongly an extended period of supply-demand balance. The scene, the researchers conclude, is set for another major oil price shock. With a chronic shortfall in supply, the world faces a permanent increase in the price of oil.

The data supplied by Petroconsultants lend support to the conclusions reached in the Oak Ridge report that deal largely with economic consequences to the United States and its standard of living in having to deal with a cartel and its pricing power.

One of the factors that has brought about the increased demand for oil — estimated to increase by more than two million barrels per day over the next seven years to more than 80 million barrels per day — is transportation. The growing demand for automobiles in China, India, South Korea and other Asian nations pinpoints the fact that oil production provides 97 percent of the fuel used in transportation. The 600 million motor vehicles worldwide will eventually consume as much as 60 percent of the world's oil.

Simply put, demand for oil is outpacing the world's ability to produce it. Nations unprepared to handle the shortfall will be paying a significantly higher price for oil.

Although wind energy is used predominantly to produce electric power, and competes primarily against gas, coal and nuclear power, its future prospects will also be affected by tighter world oil supplies and rising prices. If compressed natural gas (CNG) becomes the substitute of choice for gasoline as a transportation fuel, then gas producers, too, are likely to find themselves facing supply problems. And to the degree that public attention is focused on energy issues, a pollution-free alternative that now costs only slightly more than conventional sources can only gain.

So, there you have it. As never before, we must set our determination to both educate ourselves and take action. Learn about alternatives and renewables, apply them personally to the extent that we are able and corporately take action to get the wheels of research, development, and application moving full throttle! The learning begins here.

Citizens need to understand the problem and the alternatives that are available. This book will introduce you to the concepts and terminology dealing with energy and various alternative sources of energy to meet our ever-increasing electrical energy needs. The term alternative, as in alternative energy, is used to mean those sources of energy that are an alternative to fossil fuels. Another term that is often used is renewable, as in renewable energy sources. Renewable energy sources are those alternative energy sources that draw from inexhaustible supplies such as wind, sun, water stored in reservoirs, ocean tides, ocean currents, and ocean heat storage. Also, renewable energy sources are often referred to as *sustainable*. This book will introduce you to all of these and more. You will discover the advantages and disadvantages of each.

I wish to thank and hereby acknowledge the following people without whom this book would have been less complete. My sincere thanks goes to:

The very capable staff of PROMPT Publications of Indianapolis, Indiana, a division of Howard W. Sams & Company, including Candace M. Hall, Managing Editor, and Natalie F. Harris, Project Manager.

The individuals representing many companies and organizations who contributed time, effort, information, and photographs:

J. Hilbert Anderson, Sea Solar Power, Inc. of York, PA
William A. Beckman, Prof., Solar Energy Laboratory, U. of Wisconsin at Madison
Michael L.S. Bergey, Bergey Windpower Company, Inc. of Norman, OK
Jeffrey L. Bowden, WINDTech International, L.L.C., Atlanta, GA
Jito Coleman, New World Power Technologies of Waitsfield, VT
Charlie and Fran Collins of LaVerkin, UT
Dean Cooley, Pacific Gas & Electric Co., Geysers Power Plant, Healdsburg, CA
Corrine Cottle, FloWind Corporation of San Rafael, CA
Tom Daniel, Natural Energy Laboratory of Hawaii
Brooke L. Decker, Acuson Corporation of Mountain View, CA
Richard B. Diver, Sandia National Laboratories of Albuquerque, NM
Tony Fiore, Solaray Systems of Englewood, FL
Gail Francisco, FloWind Corporation of San Rafael, CA
Leslie L. Gardner, National Renewable Energy Laboratory of Golden, CO
John Gleason, New World Power Technologies of Waitsfield, VT
Paul A. Gottlieb, Department of Energy, Washington, DC
Tom Gray, American Wind Energy Association of Washington, DC
Adrianne N. Greenlees, Natural Energy Laboratory of Hawaii
Sarah Howell, Solarex of Frederick, MD
V. Klein, World Power Technologies of Duluth, MN
Andrew Kruse, Southwest Windpower, Inc. of Flagstaff, AZ

Becky Latson, Advanced Wind Turbines, Inc. of Seattle, WA
Jessica L. Maier, American Wind Energy Association of Washington, DC
Sandee Matarese, AstroPower, Inc. of Newark, DE
Mary McHaren, Zond Systems, Inc. of Tahachapi, CA
Robert Muhn, AstroPower, Inc. of Newark, DE
Robert Poore, Advanced Wind Turbines, Inc. of Seattle, WA
David Rippner, Alternative Energy Engineering of Redway, CA
Denise L. Scullen, Hutton Power Systems Group, Norcross, GA
Steven P. Steinhour, Kenetech Windpower, Inc. of San Francisco, CA
Steven J. Strong, Solar Design Associates of Harvard, MA
Patrick Summers, National Renewable Energy Laboratory of Golden, CO
Brett Tarnet, Siemens Solar Industries of Camarillo, CA
Anne Van Arsdall, Sandia National Laboratories of Albuquerque, NM

1

ELECTRICITY FROM MANY SOURCES

This chapter has been designed to help you understand the mechanisms that are employed in converting various forms or sources of energy into electricity. A basic understanding of these will enable you to understand more fully the alternative and renewable sources discussed in later chapters. If you are a novice to the world of electricity and energy, this chapter and the next will serve as a useful initiation. Chapter Two contains a special section titled *Basic Electricity*; a tutorial which will explain more fully concepts of voltage, current, power, and more; topics that are not covered in detail here.

As you study this book, you will discover many different forms of energy that man has learned to put to use. Usually, energy sources are harnessed for the purpose of producing electricity, which is another form of energy. To accomplish this, special devices have been devised to act as transducers. Transducers are devices that transform energy from one form to another. Each transducer type is associated with a particular source of energy. In this chapter, we will examine the most common sources of energy that can be converted to electricity through the use of special transducers. These common sources of energy involve chemistry, magnetism, heat, light, and mechanical motion. Thus, the various primary energy conversion methods are referred to as: electrochemical, electromagnetic, thermoelectric, photoelectric, and piezo-electric. Each of these is a mechanism or means by which various energy sources or types are used to produce electrical energy. In later chapters, we will discuss energy in more detail and consider a variety of systems in which various energy sources are utilized in the production of electricity. Here, we will investigate each of these primary energy conversion mechanisms, discussing theory and application. You will begin to understand the strengths and weaknesses of each as we explore their various practical applications.

1-1. THE ELECTROCHEMICAL MECHANISM

The electrochemical mechanism for creating electricity involves the conversion of energy contained in chemical reactions to electricity. In such a reaction, elements at the atomic level are forming into molecules in a process often referred to as covalent bonding. Since a detailed study of atomic structure and molecular bonding is not the purpose of this text, let me just say that covalent bonding is the means in which molecules are formed by sharing electrons in the outermost orbits (called the valence shell or highest energy level) of the atoms. In some cases during this process, electrons become free and build up a negative charge on an electrode, since each electron is negatively charged. A counter electrode in the chemical system is left with a build up of so-called positive ions (atoms that lack electrons).

What I have just described to you is what takes place in voltaic cells. A voltaic cell is a single electrochemical system that produces a specific voltage based on the chemistry involved. A combination of voltaic cells, parallel or serial, is what is technically known as a battery. However, we generally use the term battery very loosely. For example, we call flashlight cells batteries, which is technically incorrect. On the other hand, a car battery is a battery because it is a combination of six cells in series. Each cell of a standard lead-acid battery is capable of generating approximately 2.2V when fresh. We refer to the car battery as a 12-volt (12V) battery allowing for a nominal 2V per cell. Other voltaic cells of different chemistry produce different amounts of voltage per cell. Alkaline cells produce a nominal 1.5V per cell, and nickel-cadmium cells produce a nominal 1.2V. Nominal means freshly charged and under a normal load. If you are interested, you can learn more about voltaic cells from a good chemistry text or first-year college DC and AC electricity textbook. For now, let's take a closer look at the electrochemical mechanism as it exists in a standard lead-acid battery.

A typical lead-acid car battery is constructed of a set of lead-based electrodes and a bath of sulfuric acid, as shown in *Figure 1-1*. The lead plates which form the negative electrodes are made of what is called spongy lead. When sulfuric acid reacts with these spongy lead electrodes, a new chemical is formed called lead sulfate. In this reactive process, many electrons are excluded from the bonding process. These electrons build up on the negative electrodes and are available at the negative terminal of the battery. Just as sulfate is removed from the sulfuric acid to form lead sulfate, hydrogen atoms are released as positive ions (+H). These +H ions move toward the positive plates where negative oxygen ions ($-O_2$) are being formed. The positive plates are made of lead peroxide. As the sulfuric acid reacts with these plates, the $-O_2$ ions are formed. Thus, the +H and $-O_2$ ions combine to form water (H_2O). As the battery is place under load, these reaction processes are greatly accelerated and the demand for a supply of electrons is maintained and supplied to the external load (circuit) at the expense of the lead plates and sulfuric acid. Thus, as the battery discharges, what is actually happening is the chemistry is being used up. The plates and sulfuric acid are being converted to lead sulfate and water.

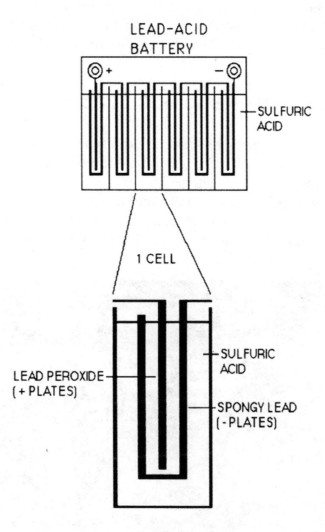

Figure 1-1. *An electrochemical system.*

When the battery is recharged, the chemical reactions are reversed by the application of an external source of electricity. In this way, electrical energy is restored to chemical energy. In the recharge process, the lead plates are rebuilt as the water and lead sulfate break down. The lead plates are renewed and the sulfuric acid is restored to a strong concentration. Since electrical energy is used to restore the battery, the battery can be thought of as an energy storage device, assuming the battery is rechargeable.

Referring back to the introduction of this chapter, you can see how the battery qualifies as an energy transducer. The electrochemical mechanism (process) is a means by which the chemical energy of the reactions is converted to electrical energy. Also, the reverse reactions during charging restores the chemical energy. We sometimes refer to such transducers as bidirectional or reciprocal.

1-2. THE ELECTROMAGNETIC MECHANISM

The electromagnetic mechanism for producing electrical energy is perhaps the most common of all. This mechanism involves electromagnetic transducers called generators (a term that is used for DC and AC transducers) and alternators (AC generators). *Figure 1-2* shows a huge steam turbine and generator system at the Palo Verde nuclear power plant near Phoenix, AZ. This installation represents many types of energy and energy conversions such as nuclear energy to heat energy; then to steam; then to mechanical motion; and finally, to electricity.

Figure 1-2. Power-plant turbines and generators are used to convert steam to mechanical energy, then to electricity. Photo courtesy of the Nuclear Energy Institute, Washington, DC.

We will investigate nuclear energy in a later chapter. A very common use of this mechanism is in our vehicles: the alternator. The alternator is belt-driven from the engine. Thus, fuel energy is converted to mechanical energy, which turns the alternator, producing electrical energy. The voltage and current produced by the alternator is AC (alternating current). Diodes acting as one-way electrical valves direct the positive alternations of the AC in one direction, and the negative alternations of the AC in another direction. In this way, the AC is converted to a pulsating DC that is regulated and used to charge the vehicle's battery. Because generators and alternators are mechanical devices, this mechanism for converting mechanical energy to electrical is often called electromechanical.

So, how does the electromagnetic or electromechanical means of converting mechanical energy to electrical energy actually work? It involves a process called induction. A voltage is induced across a coil of wire that is moving in and through a magnetic field. If a load device is connected to the generator, a current will flow. *Figure 1-3* shows an electromagnetic system for generating electricity. Let's take a moment to consider the conditions necessary to induce the voltage and current.

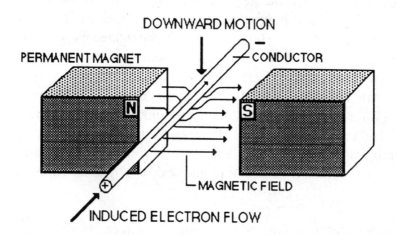

THREE THINGS NEEDED TO GENERATE ELECTRICITY
1. MAGNETIC FIELD
2. CONDUCTOR
3. RELATIVE MOTION BETWEEN THE CONDUCTOR AND MAGNETIC FIELD

Figure 1-3. A simplified electromagnetic system for generating electricity.

Three Conditions to Generate Electricity

1. First, we need a magnetic field. This magnetic field can be produced from permanent magnets or from electrical current running through a coil of wire (an electromagnet). The amount of current and voltage that can be produced by the generator is directly related to the strength of the magnetic field. The advantage to using an electromagnetic field is that the strength of the field can be varied by adjusting the current flowing through the coil, thereby controlling the output voltage of the generator. This coil is called the field coil or stator winding since it is stationary (does not move).

2. Next, and very obviously, you must have a conductor from which you can obtain an induced voltage. This conductor is part of what is called the *armature*. The armature winding has a voltage induced across it by the surrounding magnetic field of the stator. In other words, the magnetic field causes a difference of potential at the output terminals of the generator. If a load device is connected to the generator, current will flow.

3. Finally, in order for the generator to actually produce electricity, there must be relative motion between the armature and the stator windings or permanent magnet(s) if a field coil is not used. This is probably something you already knew. If the generator is not being turned, no voltage or current is produced. The armature must be rotated by some external source of mechanical energy which, of course, is derived from some other form of energy. The external energy source is then converted to electrical energy by generator action. It might be interesting to note here that, theoretically, it takes 1 horsepower (hp) of mechanical power to produce 746 watts (W) of electrical power. This does not include mechanical and electrical losses such as friction and heating effects. In practice, it is better to assume that a generator will provide approximately 600W when driven by a 1hp engine. This allows for real-world losses. As an example, a 3,000W generator should be driven by at least a 5hp engine (3,000W/ 600W/hp = 5hp).

Generators are adaptable to a great variety of energy sources. Hydroelectric power plants make use of the tremendous amount of energy stored in a lake or reservoir contained by a dam. Steam turbine generators are fired by the energy stored in coal, natural gas, heat energy released by a nuclear reaction, and the sun. The energy in the wind is harnessed to turn huge blades that are geared up to turn the armature of a generator. Even the heat stored deep in the earth and in the warm tropical oceans is being used to turn special turbine generators. We will discuss these sources of energy in later chapters.

1-3. THE THERMOELECTRIC MECHANISM

In 1821, Thomas Johann Seebeck discovered the thermoelectric mechanism, and built the first thermoelectric transducer, known as a thermocouple. In honor of Mr. Seebeck, the thermoelectric mechanism is widely known as the Seebeck effect. The thermocouple (and thermoelectric mechanism) is illustrated in *Figure 1-4*. Mr. Seebeck discovered that when two conductors made of different metals are joined on each end to form a loop, and the temperature of one end is much different than the other, a small electrical current is produced. This direct current (DC) is produced as one end of the thermocouple is cooled and the other end is heated. Unfortunately, the amount of voltage produced is very small, in the microvolts $(0.000001V = 1E - 6V = 1 \times 10^{-6}V = 1 \text{ microvolt} = 1\mu V)$, and is at an extremely low power level. The amount of voltage produced is directly related to the difference in temperature from one end to the other.

One of the most practical applications for the thermocouple is in temperature measurement devices. A very sensitive voltmeter is used to measure the electricity produced due to the difference of temperature on each end of the thermocouple. The scale of the meter is calibrated in degrees, usually Celsius or Kelvin. *Figure 1-5* illustrates a thermocouple temperature meter.

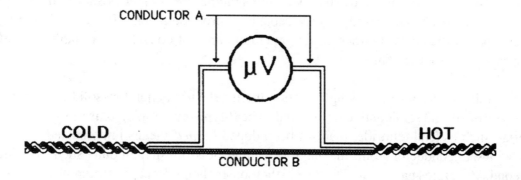

Figure 1-4. The thermocouple.

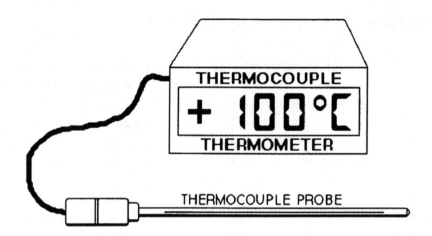

Figure 1-5. *A digital thermocouple thermometer.*

When thermocouples, or related transducers, are connected in series, more voltage is obtained; and when they are connected in parallel, more current is obtained. A series and parallel combination yields higher voltage, current, and power levels (since power is the product of voltage and current). Such a combination is known as a thermopile. A thermoelectric generator is a system in which a heat source is applied to the thermopile to produce direct current (DC) electricity. Usually, the energy conversion efficiency of thermoelectric generators is very low, ranging less than 5%. (Compare to electromagnetic generators ranging from 60% to 80%). Consequently, thermoelectric generators are rarely used as a serious source of electricity. In spite of the low efficiency, you may be interested in learning that the Apollo 12 astronauts used a small nuclear-powered thermoelectric generator to power their instruments while on the moon. More down-to-earth, I have seen some manufacturers selling small thermopiles to be installed in home fire places. These are very expensive and usually produce less than 100W of electrical power with a raging fire. On the positive side, thermopiles can be driven by any energy source capable of producing heat. Because of their high cost and poor efficiency, thermoelectric generators are not considered a viable source of electrical energy for the home.

Figure 1-6 illustrates a simple, single-section thermoelectric generator (solid-state thermopile) made of special semiconductor material. The P-type semiconductor is a semiconductor material, such as lead telluride, that has been doped (*doped* means impregnated or mixed with) with semiconductor atoms that have only three electrons in their outer shell. The main semiconductor atoms have four electrons in their outer shell. When the different atoms form bonds in a crystalline structure, only a total of seven electrons are available for the atoms to

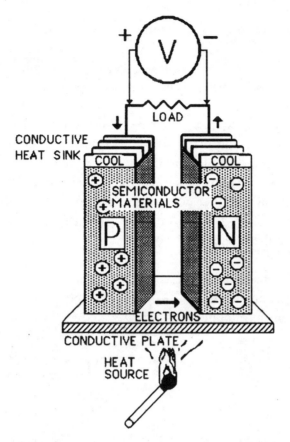

Figure 1-6. *A simple thermoelectric generator.*

share. If eight electrons were available, the bonding would be complete. In this case, the seven shared electrons fall one short of the eight, leaving a hole to which free electrons can be attracted, just as a positive potential attracts negative electrons. Thus, the overall material is referred to as P material.

N-type semiconductor material has been doped with atoms that have an excess of electrons. When bonds between the dissimilar atoms are formed, only a total of eight electrons are needed. So, one electron is rejected and becomes free. The overall material has an abundance of free electrons. Therefore, due to this abundance of free negative electrons, the material is referred to as N material.

The electrical difference between the mated N and P materials is created to produce a migration of free electrons when a difference in temperature is applied. Because of the external heat energy, or difference in external temperatures, electrical current will flow through a connected load device.

Figure 1-7 illustrates a thermopile being used as a "heat pump". When a DC current is caused to flow through the device, as shown, the upper surface will become very cold, absorbing heat and transferring it to the lower surface. If the current is reversed by reversing the battery, the top becomes hot and the lower surface becomes cold. This cooling and heating effect of a powered thermocouple was discovered in 1834 by James C. Peltier. Thus, this interesting effect is commonly known as the Peltier effect and the thermopile is referred to as a Peltier junction. Naturally, Peltier was not able to experiment with modern semiconductors since they did not exist in his time. Instead, he used dissimilar metals to make simple thermocouples and powered them with a DC source.

Modern Peltier junctions are made of N-type and P-type semiconductor materials, and have extreme temperature differentials from top to bottom, depending on the size of the junction and the amount of direct current applied. *Figure 1-7* shows the dimensions of a typical

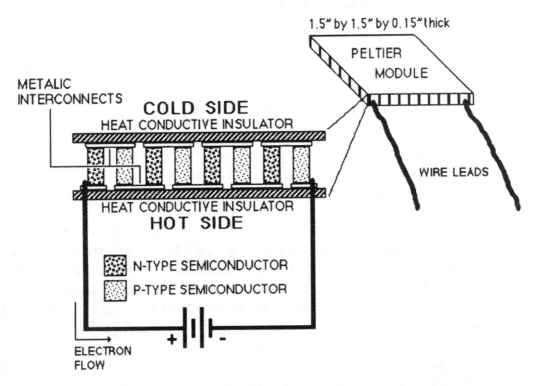

Figure 1-7. The peltier junction.

commercial Peltier junction costing approximately $30 (U.S.). Such a junction may require 12V to 15V, and draw 5 amps (A) to 10A of current. Typical applications include small picnic coolers/food warmers, and microprocessor coolers in computers.

1-4. THE PHOTOELECTRIC MECHANISM

The photoelectric mechanism for producing electricity is ubiquitous (all around) though perhaps largely unnoticed. The transducers that convert the sun's energy to electricity (DC) are often referred to as solar cells. They can be seen in many applications that require electricity in remote locations such as scientific remote environmental monitoring stations, radio relay sites, telephone and television relay sites, rural homes. They are also used to recharge batteries in remote night lighting systems. I think you've got the idea. In addition to these applications, some builders offer solar electric packages with new homes. Such packages may add $30,000 to $50,000 to the cost of the home. We'll explore this in detail later.

Like the thermoelectric generator previously discussed, solar cells must be electrically connected in series and parallel combinations to produce the desired voltage and current. For

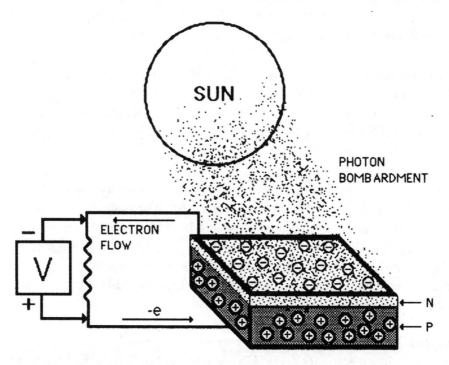

Figure 1-8. *The photovoltaic cell (PV cell).*

example, a module of cells may be configured to produce say 6V at 7A or say 18V at nearly 2.33A. In either case, the module produces approximately 42W of electricity in bright sunlight. Again, you will learn more about this in a later chapter.

The official name for a solar cell or photoelectric transducer is photovoltaic cell, also known as a PV cell. It is so named because the electrical cells produce a voltage based on a photoelectric effect. The popular theory of operation of the PV cell is illustrated in *Figure 1-8*. The PV cell is usually made of semiconductor materials based on silicon (Si). As you can see, a thin layer of N-type semiconductor material (discussed earlier) is deposited over a thicker layer of P-type material. Energy from the sun (light energy) is used to free electrons from the bonds of the P material and excite them up into the N material. This increases the difference in charge between the materials which is measured as a voltage. When a load device is connected as shown, a direct current (DC) is created from the N material, through the load, back to the P material where the electrons are driven again up into the N material, repeating the cycle. The light energy is said to be a continuous bombardment of tiny particles of energy called photons. The photons collide with the electrons, thus imparting their energy. A PV cell typically produces about 0.5V, and as high as 6A in bright sun (depending on the technology). Conversion efficiencies for photovoltaic cells range from 10 to 20 percent depending on the type of semiconductor used and type of installation. The relatively low efficiency is not an overwhelming concern since the energy source is a free one.

Solar cells, formed into modules and panels, are rated at peak output power in full sunlight in the middle of the day. The cost of the solar energy conversion is expressed as cost per peak watt. In the early 1970s, the cost was a very high $25 to $35 per peak watt (U.S.). Today, the cost is less than $5 (U.S.) per peak watt and is steadily decreasing due to new manufacturing techniques and greater demand.

Much research has been done in the area of semiconductor materials for solar cells in an attempt to increase efficiency and decrease overall cost. The traditional photovoltaic cell is made of silicon semiconductor materials. Research has led to more efficient semiconductor materials such as gallium arsenide and cadmium telluride. Some exotic solar cells, in laboratory use today, are not photovoltaic but rather electromagnetic in principle, with microscopic arrays of antennas etched in silicon, receiving electromagnetic (radio frequency) energy from the sun instead of photonic.

Many of us today are enjoying the benefits of advanced photovoltaic technology, totally unaware that part of the electrical energy delivered to our homes is solar generated. We will say more about this later.

1-5. THE PIEZOELECTRIC MECHANISM

The piezoelectric mechanism, also known as the piezoelectric effect, is one in which pressure, torque, or vibration is applied to certain crystalline minerals causing electricity to be produced. This effect was first discovered by two French scientists, Pierre and Jacques Curie, in 1880. As shown in *Figure 1-9*, they discovered that a voltage is produced between opposing surfaces of a crystal when pressure is applied. The voltage follows the rate and intensity of the pressure or applied vibration. The most common crystalline minerals that produce the piezoelectric effect are quartz, tourmaline, and rochelle salt.

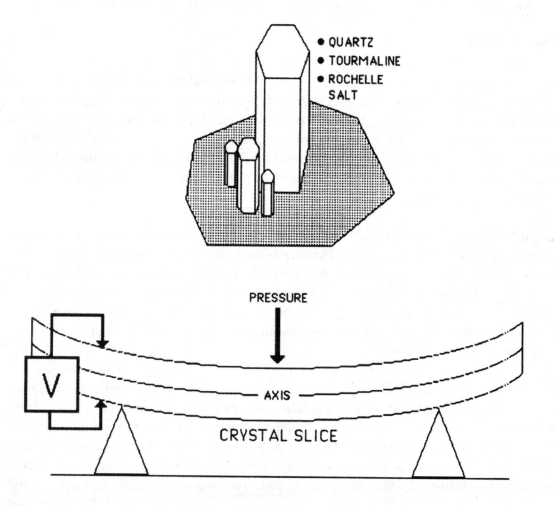

Figure 1-9. The piezoelectric effect.

The amount of voltage that can be produced at the sudden onset of pressure or a blow may range into the tens of thousands of volts. Of course, the current is extremely low, as is the power level. Though the piezoelectric mechanism is not practical to use to supply power to homes (as other mechanism may be), it is very common in many important transducers in wide use. Certain types of microphones, called crystal microphones, translate variations in air pressure to variations in electricity. In similar manner, pressure transducers convert air and mechanical pressure to voltage. Phonograph turntables, now fading into the past, had tone arms that employed a small crystal element that was tortured by the continuous vibrations of the stylus following the groove of the record. These torturous vibrations were translated into electrical patterns that were a pseudo-copy of the original audio. A discussion of many other (and more modern) applications of the piezoelectric mechanism is to follow.

Sonic Transducers

A sonic transducer is a device that is capable of converting audible sound energy (sonic) into electrical energy, and/or vice versa. A crystal microphone is an example of a piezoelectric sonic transducer. Sound pressure, as from a person's voice, causes the thin crystal element in the microphone to vibrate. The vibrations follow the frequencies and varying strength of the person's voice. These vibrations are converted to electrical energy due to the very rapid change in stress on the crystal. The electrical energy is then amplified to a useful power level for a variety of applications.

The amplitude, or amount of voltage, of this electrical signal is amazingly high (in the neighborhood of 500mVAC = 0.5V) compared to microphones of other designs (10mV to 100mVAC = 0.01V to 0.1V). However, the amount of current the crystal produces is very small. For this reason, the crystal microphone is considered a high impedance (high resistance, R) voltage source. This is demonstrated by ohm's law (R = E/I, where E is an electromotive force measured in volts, and I is intensity of flow measured in amperes). If E is high, and I is very low, R must be very high.

The crystal microphone and phono cartridge (mentioned earlier) are examples of piezoelectric transducers that convert acoustical or mechanical vibrations into electrical signals. However, piezoelectric devices are also used to convert electrical signals to mechanical vibration and acoustical energy, in a reciprocal fashion. Modern telephones no longer use a bell as a ringer for incoming calls. Instead, a piezoelectric sonic transducer is used to deliver a pulsed constant pitch tone to alert you of an incoming call. The electrical energy causes the crystal to flex or vibrate rapidly. The vibration causes an organized and continuous disturbance in air molecules, called sound waves. The physical dimensions of the crystal will determine the frequency or pitch of the audible tone.

Ultrasonic Transducers

An ultrasonic transducer is a device that converts inaudible high-frequency sound, or mechanical vibration, to an electrical voltage or signal, and/or vice versa (reciprocity). Ultrasonic frequencies are frequencies generally above 20 kilohertz (20,000 vibrations per second), outside of the range of human hearing. Piezoelectric ultrasonic transducers perform many useful functions at frequencies from 20,000 vibrations per second way up into the millions of vibrations per second (referred to as *megasonic*). Many burglar alarm systems make use of ultrasonic transducers to detect movement of an intruder (typically 40,000 vibrations per second). Usually, two transducers are used in a motion detector system. One transducer emits a continuous ultrasonic tone while a second transducer senses any changes in the returning reflected signal. The intruder is totally unaware of detection since the ultrasonic tone is inaudible to him (usually in the range of 30,000 to 50,000 vibrations per second).

One of the most interesting and perhaps most significant uses for piezoelectric ultrasonic and megasonic transducers is in the field of medicine. Today, in modern hospitals, sophisticated and accurate ultrasonic diagnostic equipment is being used to peer into the human body without exploratory surgery or dangerous radiation. Ultrasound has been used for many years to examine the fetus in a mother's womb or to detect tumors, kidney stones, or gall stones. (High-power pulsed ultrasonics is also being used to disintegrate gall stones and kidney stones, eliminating the need for surgery in many instances.) The principle involved in ultrasound diagnostics is very similar to SONAR used by ocean-going vessels. A pulse of ultrasonic or megasonic energy is sent out from the piezoelectric transducer and a returning echo is received. The time delay of the returning pulse determines the depth of the tissue and the strength of the pulse determines the density of the tissue.

Figure 1-10 and *Figure 1-11* illustrate the use of piezoelectric ultrasonic transducers in Obstetric diagnostic equipment. The ultrasound probe is often a linear (in-line) array of a hundred or more individual piezoelectric transducers with the capability of scanning and mapping a relatively large cross sectional area of the fetus. The ultrasonic echoes are analyzed and stored by a computer, and an image of a cross-section of the fetus is displayed on a television monitor. *Figure 1-10* also includes a photograph of a fetus as displayed on an ultrasound monitor. Many details are present that easily escape the untrained eye.

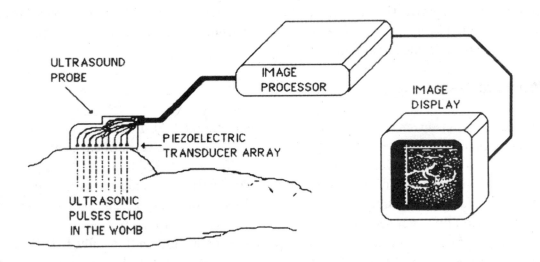

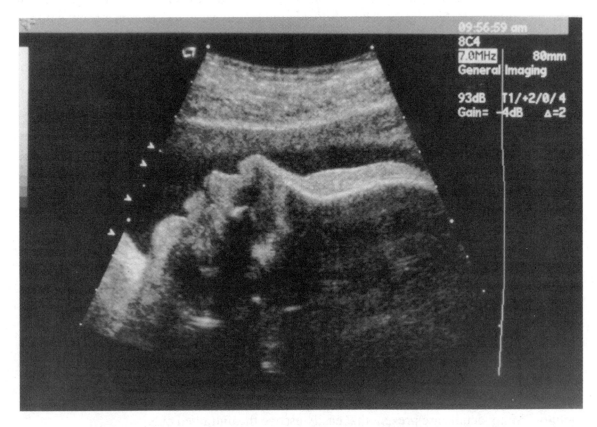

Figure 1-10. *An ultrasound scan, courtesy of Acuson, Mountain View, CA.*

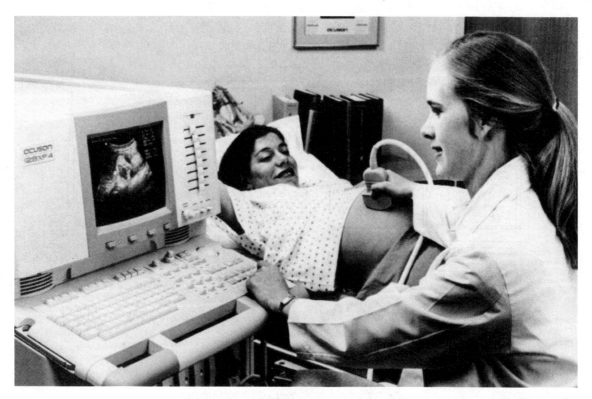

Figure 1-11. *With the new Acuson® 128XP/4 Computed Sonography System, a sonographer scans a woman in her third trimester. The XP/4 is designed for the mid-range color Doppler radiology and shared services market, offering the immediate benefits of Acuson Computed Sonography architecture while allowing a unique upgrade path to the higher performance Acuson platform. Courtesy of Acuson, Mountain View, CA.*

Radio Frequency Devices

The piezoelectric effect is also used at radio frequencies in transmitters, radio receivers, and other pieces of equipment. A thin quartz crystal in a small tin can will electrically resonate at a specific radio frequency, depending on the crystal's physical dimensions. This crystal is then used as a frequency determining element in a special circuit called an oscillator. As shown in *Figure 1-12*, these crystal-controlled oscillators are used in a great variety of equipment such as transmitters, receivers, computers, and accurate test equipment. The oscillator will produce an alternating current (AC) signal at the frequency established by the crystal. The frequency will be very accurate and stable which is very important in many applications such as communications and broadcasting.

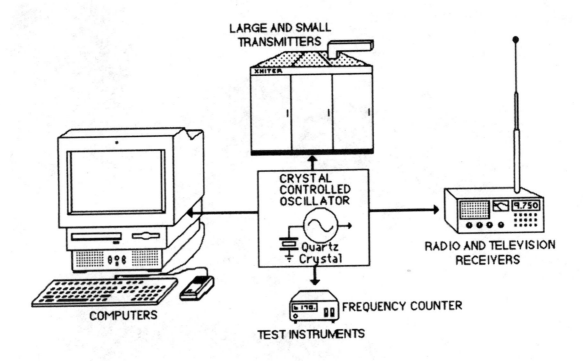

Figure 1-12. Crystal-controlled applications.

Quartz crystals are also used as special filters in radio equipment. Once again, the crystal is sensitive to a particular frequency depending on its physical dimensions. If a crystal is reso-nant at 10 MHz (10 MHz = 10 megahertz = 10 million AC cycles per second), it will allow a 10 MHz electrical signal to pass through it with little or no reduction in amplitude (voltage). Since the crystal is cut for 10 MHz, it is resonant at or sensitive to only a very narrow range of frequencies around 10 MHz. This makes the crystal a very selective radio frequency filter, passing one narrow range of frequencies and rejecting or blocking all others. We will say much more about this in a later chapter on filters.

The piezoelectric effect is an interesting one that has many applications in technology. We have covered only a few of its many areas and applications in an attempt to introduce this fascinating field. It is not surprising that many physicists and electronics engineers are devot-ing lifetimes to its study and implementation.

In this chapter, we have examined the five main mechanisms by which electrical energy is produced from other sources or types of energy. You should not only be aware of them but you should understand that some form of energy is needed to act upon each of these mechanisms before any electricity can be produced. In the following chapters, we will explore the concept of energy and some alternative sources of energy. You may wish to take a moment now to test your understanding of these mechanisms by answering the following questions. If you are on the Internet, you will want to visit some of the web sites listed in Section 1-7.

1-6. REVIEW QUESTIONS

1. Name the five common sources of electricity.
2. Give an example of an electrochemical source of electricity.
3. List the three requirements for generating electricity electromagnetically.
4. What is a thermopile?
5. Is the photoelectric source of electricity used to produce electricity for homes etc.?
6. Briefly explain the piezoelectric effect.
7. What is the difference between a sonic and an ultrasonic transducer?
8. List two applications for ultrasonic transducers.

Check your answers in the *Answers to Questions* appendix at the back of this book.

1-7. WORLD-WIDE WEB SITES

The Author's Home Page

http://www.castlegate.net/personals/hazen/

Relating to Electrochemistry

Central Electrochemical Research Institute (CECRI), India
    http://chpc06.ch.unito.it/electrochemistry.html

The Electrochemistry Gateway
    http://www.soton.ac.uk/~slt1/EchemGate.html

Relating to Electromagnetic

GE Power Systems - Generators... An Overview
    http://www.ge.com/geps/turbines/gen-hst.html

Relating to Thermoelectric

Thermoelectric Cooling (TEC)
http://www.trademart.com/aoc/tec.htm

Teledyne Brown Engineering - Hydrogen and Thermoelectric Generators
http://www.tbe.com/products/generators/generators.html

Relating to Photoelectric

Sandia National Laboratories - Photovoltaic Technology
http://www.sandia.gov/Renewable_Energy/photovoltaic/pv.html

NW Solar Cell
http://www.seanet.com/Users/miknel/NW-SolarCell.html

Centre for Photovoltaic Devices and Systems
http://www.vast.unsw.edu.au/pv.html

Relating to Piezoelectric

High-Temperature Pressure Transducers
http://www.ssec.honeywell.com/papers/HTPT/HTPT.html

Teledyne Brown Engineering - Transducers and Sensors
http://www.tbe.com/products/sensors/sensors.html

NASA - Instrument and Sensing Technology
http://ranier.oact.hq.nasa.gov/sensors_page/insthp.html

2

ENERGY

From where do we get electrical energy? Do you think you know? If you are thinking that electrical energy comes from photovoltaic cells, generators, crystals, thermocouples, and batteries, you need to think again. Electrical energy does not come from these or any other devices: electrical energy comes from other forms of energy! It is important to clearly understand the difference here. Devices (transducers) are merely the means by which various forms of energy are converted to electrical energy (or vice versa). The device itself cannot create energy. Energy cannot be created and neither can it be destroyed. It may only be converted from one form to another (the Law of Conservation of Energy). There is already more energy in our solar system alone than mankind will ever be able to use. The problem is, and always has been, to efficiently convert this energy into practical, useful forms such as electricity. So, from where does electrical energy come? It comes from as many sources of energy as mankind has been able to successfully transform. Surely, it comes from a clear understanding of the laws of physics and the nature of energy itself. In this chapter, we will take a brief look at the concept of energy, different forms of energy, how energy is measured, and a short tutorial on basic electricity.

2-1. ENERGY DEFINED

The word "energy" comes from the Greek word *energeia,* which means "activity." It carries the connotation of doing work and is very simply defined as the capacity for doing work. It is a very practical definition, one that we may quickly relate to. How often have you heard someone say, "I just can't do another thing — my energy is gone"? That person is actually saying that his or her capacity for doing work has diminished due to the amount of work that has already been done. The act of doing work, or expelling energy, is the act of converting energy to other forms. Though it is true that the overall quantity of energy is conserved, it is not true that it be conserved in the same form.

2-2. FORMS OF ENERGY

Energy is found to exist in many basic forms: potential, kinetic, thermal, chemical, nuclear and electrical. Electrical energy is obtained from other forms of energy. It would be ideal to convert directly to electrical energy from other forms of energy in hopes of obtaining a conversion efficiency of 100 %. In reality, conversion to electrical energy is far from 100% efficient. The process involves conversion to other forms of energy in addition to electrical energy, such as heat and electromagnetic radiation. As a result, conversion efficiency is less than ideal. While it is true that 100% of the source energy is converted, it is not true that it is converted 100% to electrical energy (conversion losses).

Potential Energy

As an example of what has just been discussed, consider a hydroelectric power plant. The source of energy, to be used in conversion, is the massive body of water in an elevated reservoir. This massive body of water represents potential energy. Potential energy is relational energy. That is, it is energy that is held by a stationary mass that is being acted upon by a force in relationship to a certain point. In this case, gravity is acting on the water relative to the turbines at the base of the dam. The tremendous amount of water pressure, created by the difference in level from the surface of the water to the turbine below, constitutes a great amount of potential energy.

Kinetic Energy

When the high pressure gates are opened and the water is permitted to flow, some of the potential energy is converted into kinetic energy. The word "kinetic" comes from the Greek word *kinetikos*, which means "for putting in motion." Kinetic energy is the energy contained in a moving mass. In this case, the kinetic energy is often referred to as hydraulic energy since it is water (fluid) in motion. The hydraulic energy is then converted to another form of kinetic energy, called mechanical energy, as the pressure of the water forces the turbines to rotate.

Thermal Energy

Some of this kinetic energy is converted into thermal energy due to friction in the bearings on the drive shaft and generator. Thermal energy is energy produced as a result of friction, chemical reaction, or nuclear reaction, and is characterized by a change in temperature. Finally, kinetic energy is converted into electrical energy but not without further thermal energy loss due to resistance in conductors and imperfect core material in the generator. The overall conversion efficiency for the typical hydroelectric plant is in the range of 65 to 75%.

Chemical Energy

Another common form and source of energy is chemical energy. Chemical energy is the energy contained in the molecular structure of a substance. It is actually a form of potential energy which is converted to other forms of energy through some type of chemical reaction. In most cases, chemical energy is converted to thermal energy such as in the burning of coal and petroleum products. As you know, electrical energy, along with some thermal energy, is the result of the chemical reaction in batteries. Chemical energy has been and will continue to be a valuable source of energy for conversion to electricity.

Nuclear Energy

Nuclear energy is the potential energy contained in the nucleus of an atom. It is the binding energy of the neutrons and protons. The energy contained within the nucleus of an atom is far greater than the energy that binds the electrons to the nucleus or that binds one atom to another. Thus, energy that is released in a chemical reaction (dealing with electron bonds) is puny compared to the energy that is released in a nuclear reaction (nuclear bonds). As an example, one atom of uranium will yield approximately 3.2×10^{-11} (0.000000000032) joules of energy in a nuclear reaction as compared to only 6.4×10^{-19} (0.00000000000000000064) joules of energy released by a carbon atom during chemical reaction (rapid oxidation, combustion). The uranium nucleus releases approximately 50 million (50,000,000) times more energy in a nuclear reaction than does the carbon atom during rapid oxidation. To emphasize this point once more, the energy released by 1 gram of uranium is millions of times greater than that which is released by 1 gram of combusted gasoline. Naturally, this has made nuclear energy a very attractive source of energy for the production of electricity. We will discuss this further in the section covering nuclear energy conversion.

2-3. MEASURING ENERGY

Energy is usually measured in terms of joules (J), British thermal units (BTU), or kilowatt-hours (kWh). Any of these units may be used to express a quantity of energy for any energy source since conversion between units is mathematically possible. Thermal energy is expressed in BTUs. One BTU is the amount of energy needed to raise the temperature of 1 pound of water 1° F. Normally, electrical energy is expressed in kilowatt-hours or joules, since the joule is electrically defined as one watt-second ($1J = 1W \times 1s$). Various types of mechanical energy may be expressed in joules since the joule is also defined as the amount of work done by one Newton of force acting through a distance of one meter. One of the most common units of mechanical power is the horsepower (hp). Since power used over a period of time represents energy usage, the horsepower-hour (hph) can represent a quantity of mechanical energy.

Any mathematical conversion from one unit of energy to another assumes a conversion efficiency of 100%. For example, 1hph = 0.746 kWh = 746 watthour (Wh) only if the conversion from mechanical to electrical energy is complete. In practice, this is not so, and the efficiency factor of the conversion system must be known. *Table 2-1* contains conversion formulas for various forms of energy and assumes a conversion efficiency of 100%.

As an example, suppose you want to convert 55,000 BTU to kWh. Since there are 0.000293 kWh per BTU, the number of BTUs is multiplied by 0.000293. In this case, 55,000 BTU x 0.000293 kWh/BTU = 16.115 kWh.

In the following chapters, we will discuss some of the energy sources that mankind is learning to tap as alternatives to the diminishing coal and petroleum energy sources that have, in the past, been so readily available. If you are not familiar with electricity in terms of basic theory, terminology, and practical calculations, the next section has been prepared for you. It will help you appreciate more fully the chapters that follow.

UNIT	Multiply By	To Get
joule	0.000948	BTU
joule	2.77 E-7	kWh
joule	3.72 E-7	hph
BTU	1055	joules
BTU	0.000293	kWh
BTU	0.000393	hph
kWh	3412	BTU
kWh	1.341	hph
kWh	3.6 E6	kWh
hph	2544.4	BTU
hph	0.746	kWh
hph	2.685 E6	joules

Table 2-1. Conversion formulas for various forms of energy.
100% conversion efficiency is assumed.

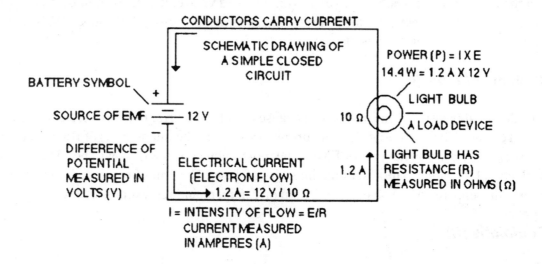

Figure 2-1. Basic electrical concepts.

2-4. BASIC ELECTRICITY: A TUTORIAL

If you are not familiar with electricity, basic theory, terminology, and basic calculations, then this section is just for you. This section is designed as a quick-start tutorial to bring you into the world of electricity quickly without a lot of rabbit-chasing and double-talk. *Figure 2-1* illustrates what we will discuss here.

Voltage (V)

Voltage is the electrical pressure that makes electricity flow in a circuit. It is often referred to as a difference in potential between two terminals, like battery terminals. On one terminal, there is a surplus of negatively charged electrons creating a negative potential. On the other terminal, there is a deficiency (lack) of electrons leaving the terminal positively charged creating a positive potential. The unit of measure for electrical potential is voltage (V). If the negative terminal has a potential of -6V (negative six volts) and the positive terminal has a potential of +6V (positive six volts), the overall difference of potential is 12V.

In addition to referring to voltage as a difference of potential, it is often referred to as an electromotive force (EMF, or just E). EMF is the potential difference acting as a force which motivates the electrons to flow. We can say that a car battery has a nominal EMF of 12V and that the difference of potential between terminals is 12V. In electrical formulas, the E is used as the symbol to represent voltage. Sometimes a V is used instead of the E. We'll look at some basic formulas in a moment.

Intensity of Flow (I)

The electrical pressure measured in volts causes electrical current to flow in a completed or closed circuit. The amount of flow of current is known as the intensity (I) of flow. The more electrical pressure you have (E or EMF), the more intense (I) the flow will be. The I is used in electrical formulas to represent the intensity of flow. The unit of measure for intensity of flow is the ampere (A). You can say 1 amp or 1 ampere and represent it as 1A.

Coulombs (C)

The flow of current is made up of billions of electrons moving through the circuit. In fact, 1A of current is equal to 6.25×10^{18} electrons passing a given point in a circuit every second. That's a lot of electrons: 6,250,000,000,000,000,000 electrons every second! This huge quantity of electrons is referred to as a coulomb (C). Therefore, 1A of current is equal to 1C of electrons passing a given point in a circuit every second (1A = 1C/1s). 5A would be 5C per second, and so on.

Resistance (R)

The opposition to the flow of electrical current is known as resistance (R). The resistance in a circuit limits the amount of current that will flow. Resistance is usually good and needed to control current. The more resistance there is in a circuit, the less the current or intensity of flow. In other words, the more R there is, the less I there will be. The R is used in electrical formulas to represent resistance. The unit of measure for resistance is the ohm (Ω, the omega symbol). We write it like this: one ohm is 1Ω.

Ohm's Law (I = E/R)

Ohm's Law states that the intensity of flow (I) is directly related to the amount of electromotive force (E or EMF) and inversely related to the amount of resistance (R) in a circuit. For example, if 12V of electromotive force is applied to a circuit having a total of 6Ω of resistance, the intensity of flow will be 2A ($I = E/R = 12V/6\Omega = 2A$). If the EMF is increased to

24V, and circuit resistance is not changed, the resulting intensity of flow will be 4A (I = E/R = 24V/6Ω = 4A). If we then increase the resistance to 12Ω, the intensity of flow will return to the original 2A (I = E/R = 24V/12Ω = 2A).

The Ohm's Law formula can be rearranged using basic algebra to solve for R when E and I are known. R = E/I. Resistance is equal to electromotive force divided by intensity of flow. For example, if we measure a current of 3A (using an ammeter) flowing in a circuit to which there is applied an EMF of 12V, we can calculate the amount of circuit resistance to be 4Ω (R = E/I = 12V/3A = 4Ω).

Again, we can rearrange the Ohm's Law formula to solve for electromotive force (E) when resistance (R) and intensity of flow (I) are known. E = I x R, which can also be written E = I • R or E = IR. As an example, let's say a circuit is known to have a total resistance of 10Ω and a current of 2A. The applied electromotive force (E or EMF) must be 20V (E = I x R = 2A x 10Ω = 20V).

Power (P)

Power is the rate at which energy is being used or converted to other forms of energy. The unit of measure for electrical power (P) is the watt (W). One watt (1W) of electrical power is the amount of power converted when one joule (1J) of energy is expended every second of time (1W = 1J/1s). If 10J of energy is expended every second, 10W of power is being used. The higher the power in watts, the faster the energy is being consumed or converted to another form of energy such as heat, light, and/or radiation.

The amount of power being used by a circuit or individual component of the circuit can be calculated in any of three ways: P = I x E (the pie formula), P = I^2 x R, and P = E^2/R. Consider the following three examples:

1. How much power does a circuit dissipate (convert to heat, light, radiation, etc.) if the current is measured to be 5A and the applied voltage is 120V?
   Answer: P = I x E = 5A x 120V = 600W.

2. How much power does the wiring in an electrical circuit waste if there is 20A of current and the resistance of the wire is 0.1Ω?
   Answer: P = I^2 x R = 20A^2 x 0.1Ω = 400 x 0.1 = 40W.

3. How much power does an individual 5W resistor in a circuit dissipate if the voltage across the 5W resistor is measured and found to be 20V?
   Answer: $P = E^2/R = 20V^2/5\Omega = 400/5 = 80W$.

Electrical Energy Consumption (kWh)

Accumulated energy usage is what we pay for when our electric bill comes. It is measured in kilowatt-hours (kWh) which is power times time. For example, 1 kWh = 1 kW x 1 hour, 5 kWh = 1 kW x 5 hours, 20 kWh = 2 kW x 10 hours, etc. As a practical example, let's say your 5,000W hot-water heater runs a total of 6 hours per day. The total electrical energy consumption for one day is 30 kWh (5 kW x 6 hours = 30 kWh). Note: 5,000W = 5 kilowatts = 5 kW. For a 30-day month, the total energy consumption will be 900 kWh (30 kWh/day x 30 days = 900 kWh). Now, let's say you pay 10 cents for every kilowatt-hour of energy usage. Your total cost for heating water for one 30-day month is $90 (900 kWh x $0.10 = $90). If $90 sounds like a lot to you, you can consider installing a solar water heating system or install a $40 hot-water heater timer. Sometimes power companies will actually pay for this to be done for you. You'll find that an hour early morning and an hour late afternoon will probably be adequate to meet your hot water needs. You will have 2 hours per day usage instead of 6. Actually, you will have less than that because it will not take the heater a full hour to heat your water during each 1 hour on period. So, let's say the heater is actually heating water for 1.5 hours. You have just cut your monthly bill down to $22.50 instead of $90. Not bad!

DC and AC

You might be wondering what the difference is between DC and AC. First, DC stands for direct current and AC stands for alternating current. DC is electricity that always flows in the same direction. A battery-powered circuit is a DC circuit since electricity always flows from the negative terminal of the battery, through the circuit, back to the positive terminal of the battery (electron flow). With AC, current reverses direction continuously. This is because the voltage applied to the AC circuit is continually changing polarity. Each current direction is called an *alternation*. Two alternations make up a cycle. The time (T) for one cycle is the time is takes to complete one positive alternation and one negative alternation. The number of cycles that are repeated every second is known as the frequency (f) of the AC. The relationship between cycle time and frequency is simply an inverse relationship where f = 1/T and T = 1/f. For example, in the U.S. and Canada, the AC line frequency is 60 cycles per

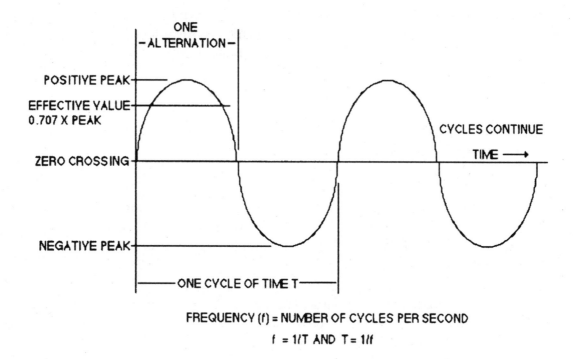

Figure 2-2. *The AC waveform and calculations.*

second (60 cps = 60 Hertz = 60 Hz). That means for every second there are 60 cycles and 120 alternations. The time for one cycle is simply f = 1/T = 1/60 = 0.0167s = 16.7 millisec-onds = 16.7 ms.

Is AC voltage and current different than DC voltage and current? Yes. AC voltage and current rise to a peak in one direction then fall back to zero, then continue and rise to a peak in the opposite direction. This is illustrated in *Figure 2-2*. The positive and negative peak values of AC voltage in the U.S. are about 170V (+170Vp and -170Vp). This peak value is short-lived and does little work. It is an overall effect of the alternation that is important. A 170Vp alternation has the same effect as 120VDC. That's why we usually always refer to AC voltages as effective values, not peak. So when you consider 120VAC, it is actually the effective value and has the same effect as 120VDC. Converting an AC peak voltage to an effective value is done by simply multiplying the peak value by the constant 0.707. In other words, 170Vp x 0.707 = 120V. To convert an effective value to peak, simply divide the effective value by 0.707 like this: 120V/0.707 = 170Vp. All calculations using ohm's law and the power formulas are usually done using the effective AC values of voltage and current. You will see this in the following household applications.

Common Applications

1. How much current will a 100W light bulb draw if the voltage is 120V?
    Answer: $P = I \times E$. Therefore, we can rearrange this formula using basic algebra to solve for I. $I = P/E = 100W/120V = 0.833A$.

2. A microwave oven consumes 1200W of electricity while operating at full power from a 120V circuit. What is the equivalent internal resistance of the microwave oven?
    Answer: $P = E^2/R$. This formula can be rearranged to solve for R: $R = E^2/P = 120V^2/1200W = 14400/1200 = 12\Omega$.

3. A certain photovoltaic module can provide 18V and a maximum of 5A in bright direct sunlight. How much power can it provide?
    Answer: $P = I \times E = 5A \times 18V = 90W$.

4. A color TV draws 1.5A of current from a 120V outlet. How much power does it consume?
    Answer: $P = I \times E = 1.5A \times 120V = 180W$.

5. An air conditioning system draws 14 A of current from a 240 V circuit. How much power is it consuming?
    Answer: $P = I \times E = 14A \times 240V = 3,360W$.

6. If the audio voltage applied to an 8Ω speaker is doubled from 10V to 20V, what is the increase in audio power?
    Answer: The audio power at $10V = E^2/R = 10V^2/8\Omega = 12.5W$. The audio power at $20V = 20V^2/8\Omega = 50W$. As you can see, when voltage is doubled, the power is quadrupled. That's because the formula includes E^2.

7. Calculate the power loss in a 1000-foot length of wire that has a total resistance of 0.15Ω while 30A of current flows.
    Answer: $P = I^2 \times R = 30A^2 \times 0.1\Omega = 90W$.

How much voltage is there dropped across this wire?
Answer: E = I x R = 30A x 0.1Ω = 3V.

If 120V is applied to the wire and the wire drops 3V, how much voltage is left for the load at the far end?
Answer: 120V - 3V = 117V.

How much power does the load device actually receive?
Answer: P = I x E = 30A x 117V = 3510W.

How much power was there supplied from the 120V source?
Answer: P = I x E = 30A x 120V = 3600W.

Calculate the difference between the input source power and the load power.
Answer: $P_{diff.} = P_{loss} = P_{source} - P_{load}$ = 3600W - 3510W = 90W.

8. A hot-water heater is connected to a 240V supply and consumes 5,000W of power when it is on. How much current does the water heater demand when it is on?
Answer: P = I x E. Therefore, we can rearrange this formula to solve for I. I = P/E = 5,000W/240V = 20.8A.

9. If you pay 10 cents per kWh for electricity, how much will it cost you to use your 1200W microwave oven every month if it is in use about 1 hour per day? Note: 1200W = 1.2 kW.
Answer: 1.2 kW x 1 hour/day x 30 days/month x $0.10/kWh = $3.60/month.

10. How many kWh of energy can you collect in one day if your photovoltaic array (solar electricity array) produces 18V at 50A for 6 hours?
Answer: P = I x E = 50A x 18V = 900W, and kWh = 900 W x 6 hours = 5400 Wh = 5.4 kWh.

2-5. REVIEW QUESTIONS

1. From where does electrical energy come?
2. What is the Law of Conservation of energy?
3. Define energy.
4. What is kinetic energy?
5. How many joules and kilowatt-hours of energy are there contained in 0.45 BTU?

6. Calculate the amount of current flowing in a closed circuit that has a total resistance of 50Ω and an applied voltage of 24V.
7. How much power will a 10Ω resistor dissipate if the current flowing through it is 4A?
8. Determine the accumulated electrical energy usage of a 100W light bulb used for 4.5 hours.
9. Calculate the monthly cost of operating a 2kW machine for 16 hours per day for 20 days if the cost rate is 12 cents per kWh.
10. If an AC voltage has a peak of 340Vp, what is its effective value?

Check your answers in the *Answers to Questions* appendix at the back of this book.

2-6. WORLD-WIDE WEB SITES

The Author's Home Page

http://www.castlegate.net/personals/hazen/

Relating to Energy

California Energy Commission
 http://www.energy.ca.gov/energy/html/directory.html

Energy Story
 http://www.energy.ca.gov/energy/education/story/story-html/story.html

Energy Yellow Pages
 http://www.ccnet.com/~nep/yellow.htm

National Renewable Energy Laboratory (NREL)
 http://www.nrel.gov/

Relating to Power

University of California Energy Institute
 http://www.ucenergy.eecs.berkeley.edu/ucenergy

Sandia National Laboratories
   http://www.sandia.gov/

U.S. Department of Energy (DOE)
   http://www.doe.gov/

3

SOLAR ENERGY

It is well-known that our greatest source of energy is the sun. The sun is one enormous and almost perpetual nuclear reaction that conveys vast amounts of energy (approximately 2.1 x 10^{15} kWh per day) to us in the form of *electromagnetic radiation*. *Electromagnetic energy* is energy that is contained in x-rays, gamma rays, light, and lower frequency radio waves. It is converted from other forms and sources of energy either naturally, as from the nuclear reactions on the sun, or through man-made devices such as light and heat sources, transmitters, and nuclear reactors. *Electromagnetic radiation* is composed of electric energy waves (having to do with a force-field created by a difference in potential) and perpendicular magnetic energy waves. These energy waves have the ability to transport electrical and thermal energy over great distances.

It is interesting to note that many of the energy sources that we use to produce electricity, actually originate from the sun and, therefore, could be broadly considered as forms of solar energy. Hydro energy, as used in hydroelectric plants, is a form of solar energy because the sun evaporates the water that causes the rain which fills the reservoirs. Solar heating causes winds which turn the blades on wind powered generating systems. Thermal energy stored in tropical oceans comes directly from the sun. The chemical energy stored in coal and petroleum products that fire our steam turbine generating systems was mothered by the sun through photosynthesis ages ago.

As you will see, the topic of solar energy is a major one covering many areas of technology. In this chapter, we will concentrate on two modern and major areas of solar technology that are being used to generate electricity: *solar thermal* and *photovoltaic (PV)*. The area of solar thermal technology is very broad, covering solar space heating and solar water heating (as substitutes for electrical usage), as well as ocean thermal, solar super heating, and steam generation for the production of electricity. The photovoltaic area of solar energy technology, on the other hand, deals with direct conversion of the sun's rays to electrical energy. In the discussions to follow, we will concentrate on those areas of solar energy that are related to the production of electricity.

3-1. PHOTOVOLTAIC (PV) ELECTRICITY

Great strides are being made in the area of photovoltaic technology as researchers are discovering new ways to convert the sun's rays to electricity at "reasonable" costs. The efficiency of PV cells has crept up from less than 10% in early designs to present efficiencies ranging up to about 30% for elaborate high-technology laboratory installations involving special lenses, filters, and cooling systems. Most practical installations in operation around the world today are very basic in design, and operate at about 11% conversion efficiency. It may sound inefficient, but remember the energy source is free and inexhaustible (renewable).

Power From PVs

So, how much energy does an 11% efficiency represent? The sun's energy is collected and converted by solar panels with a certain surface area. Naturally, larger surface areas will collect more energy. According to the National Aeronautics & Space Administration (NASA), approximately 1353W (watts) of solar power per square meter (1353 W/m^2) is available outside of the earth's atmosphere. Approximately 10 to 30% of this power is lost due to reflection or absorption in the atmosphere by air, dust, and moisture, assuming a clear sky. This means that between 950Wp/m^2 and 1220Wp/m^2 actually strike the earth on a clear day. The amount of solar power impinging on a square meter of earth's surface is known as the *insolation level*, and is expressed in watts/meter. Watts peak (Wp) occurs when the sun is directly overhead and perpendicular to the receiving surface. Thus, a small PV array, with a surface area of 1m^2, will receive an average of 1100Wp of solar power and deliver about 121W of peak electrical power (121Wp), assuming 11% efficiency. For example, a 5m^2 PV array would deliver about 605Wp (121Wp/m^2 x 5m^2 = 605Wp). This, of course, is based on 1200Wp/m^2. Many designers use a more conservative figure of 1000Wp/m^2 for calculations.

PV Technologies: Types of Cells

Basic Theory

A discussion of various PV technologies involves an investigation into semiconductor technology similar to that which we began in Chapter 1, as we discussed the photovoltaic cell. Recall that the photovoltaic cell is an electrical transducer that is able to convert light energy (photon bombardment) into electrical energy, producing a flow of electrons when a load device is connected. As we discussed, the PV cell is made primarily of a semiconductor material called silicon (Si). We learned that the addition of impurities (dopants) creates either N-type material or P-type material. Traditionally, phosphorous has been used as a dopant to produce N-type material and boron to produce P-type.

The N-type material is rich in free or mobile electrons. The N stands for negative since each of these free electrons is negatively charged. However, since the overall number of electrons, free or captive, is balanced by an equal number of protons (positive particles in the nuclei of the atoms), the net (overall) charge of the N-type material is neutral.

With P-type material, the bond structure of silicon and boron atoms actually produces "holes" — blank places where an electron should be, and would be, if one was available. Thus, in a sense, the bond structure of silicon is lacking electrons, or lacking negative charges. Thus, the material is called P-type for positive. However, once again, the net charge of the P-type material is still neutral since the actual number of negative electrons forming bonds is balanced by an equal number of positive protons in the nuclei of the atoms.

If no light is applied to the PV cell, made of P- and N-type material layers, there will remain a net neutral charge on each material (layer). (See *Figure 1-8.*) Therefore, there is no difference in charge (voltage) between the two layers of differing materials; thus, no electrical energy is produced and a load cannot be driven. When light is applied (photon bombardment), the photons quickly pass through the very thin N-type layer and enter the P-type material below. Large numbers of electrons in the P-type material are freed from their bonds due to the photon collisions. In other words, the impacted electrons are given a boost of energy which allows them to move up into the N-type material, creating an even greater abundance of free electrons in that material. Because of this forced migration of electrons, the P-type material is actually left with a net positive charge, and the N-type material gains a net negative charge. Thus, one layer has a negative charge while the other has a positive charge, and the difference between the two charges is a usable voltage that can drive an electron flow through a load if one is connected. Typically, a single cell produces a peak voltage in the neighborhood of 0.5V. The amount of current and power the cell produces is related to the size of the cell (surface dimensions facing the light source). As an example, a 10 cm. (~4") square single-crystal silicon cell in bright sun will produce almost 0.5V and provide a current of nearly 3A.

Single-Crystal Silicon (x-Si) Cells
The selection of PV technologies available today vary greatly. Basically, they all function according to the theory just discussed. However, they vary widely in manufacturing cost, conversion efficiency, and manufacturing processes. One of the oldest and most expensive, but most efficient technologies is known as single-crystal silicon (x-Si) technology. Single-crystal silicon (x-Si) is silicon material that is "grown" in a high-temperature oven, forming cylinder-shaped ingots called *boules*.

How Silicon Boule is Made

Chunks of pure silicon are placed in a quartz crucible (bowl) which is then placed in a furnace. The silicon is melted at a temperature of approximately 1500° C. Within the furnace, a thin silicon crystal "seed," roughly 15 cm long (6"), is lowered into the molten silicon. As the silicon cools, the seed is rotated slowly in the melt until an ingot of the desired diameter is formed. Then, the formed boule is slowly raised out of the cooling melt, and the length of the cylindrical boule continues to increase. The entire process is carefully controlled by a programmed computer.

The ingot is cut into very thin circular slices called blanks, a process that usually wastes as much silicon as is spared. Some manufacturers cut the sides of the boule before slicing to form a rectangular brick, which yields square slices when cut. (This increases sun exposure area when the cells are placed together to form a module, thus increasing conversion efficiency.) The slices are then doped in a high-temperature process in which the impurities (dopants) migrate (diffuse) into the silicon material. All of these manufacturing steps are expensive and wasteful, causing the unit price of a finished cell to be relatively high. However, conversion efficiency can be as high as 20%. Long-term stability of these cells make them very desirable where initial cost is not a primary consideration.

Ribbon-Silicon Cells

Related to the x-Si cells discussed above are the ribbon-silicon cells. A ribbon cell is illustrated in *Figure 3-1* along with the sliced x-Si cells. A ribbon cell is a single-crystal silicon cell that is manufactured in a process similar to metal extrusion. Molten silicon seeps up through a thin yet wide channel by capillary action, and is drawn out slowly by mechanical means, forming a continuous long ribbon of thin silicon. Advanced manufacturing techniques enable multiple ribbons or hollow, multisided tubes to be pulled at the same time. The sides are then cut to form ribbons. The ribbons are cut, usually by laser, into desired lengths. Additional high-temperature processing includes the diffusion of dopants into the ribbon to produce the N- and P-type layers and metal deposition, to form the charge collector conductors. The individual cells can then be laid into module frames of the desired size — series interconnections for more voltage and parallel interconnections for more current. Though these cells are single-crystal silicon (x-Si), their energy conversion efficiency is not as high as the sliced kind. The efficiency for ribbon cells is usually in the range of 10% to 14%, mostly closer to 10%.

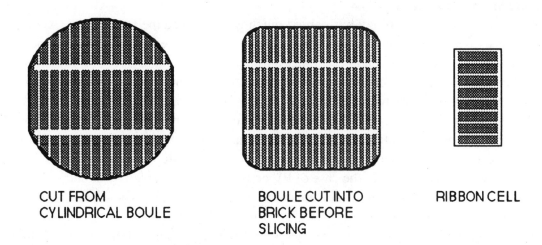

CUT FROM
CYLINDRICAL BOULE

BOULE CUT INTO
BRICK BEFORE
SLICING

RIBBON CELL

Figure 3-1. Single-crystal silicon PV cells.

Polycrystalline-Silicon (P-Si) Cells

Figure 3-2 illustrates the characteristic appearance of polycrystalline-silicon (P-Si) PV cells. Notice the random yet obvious crystalline structures formed throughout the slice. This kind of cell is less costly to manufacture and has a somewhat lower conversion efficiency than single-crystal silicon cells. To increase the efficiency of P-Si, the cell is hydrogenated. That means the cell is impregnated with hydrogen atoms to improve electron mobility through the cell. This is done by placing the P-Si material in a vacuum chamber and bombarding it with hydrogen atoms, or ions. The result is a solar cell manufactured at a lower cost than x-Si cells, which has a slightly lower conversion efficiency, usually 12% or 13%.

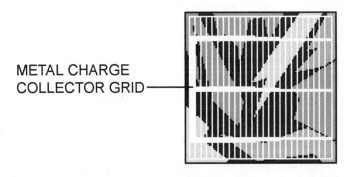

METAL CHARGE
COLLECTOR GRID

Figure 3-2. A polycrystalline silicon PV cell.

Amorphous-Silicon (a-Si) Cells

Amorphous-silicon technology has gained much ground since the mid 1980s. Amorphous silicon is used in many applications today ranging from flat-panel displays (such as the liquid-crystal display screens on laptop computers) to photovoltaic cells. It is often referred to as thin-film technology. In thin-film processes, thin doped or undoped silicon layers are deposited on sheets of glass, metal, or another material. This is done through a process known as *chemical vapor deposition* (CVD). Silicon and dopants are introduced in gaseous form (silane, phosphine, etc.) into a special chamber where relatively low-temperature ($250°$ C to $500°$ C) reactions take place that result in an even layer buildup of silicon on the substrate (glass, stainless steel, etc.). In the case of PV cells, multiple layers of N-doped silicon, pure (intrinsic) silicon, and P-doped silicon are deposited consecutively. The layers are referred to as "thin film" because they are only 20 to 5,000 angstroms thick. One angstrom ($Å$) is 1×10^{-10} meters thin. A sheet of 20 lb. copier paper is approximately one million angstroms thick! A human hair is between 500,000$Å$ and 700,000$Å$! It helps you understand just how thin only 20$Å$ is. Very thin!

Amorphous silicon is not usually anywhere near as efficient as single-crystal or polycrystalline silicon. Charge mobility is normally a thousand times or more less in a-Si than in x-Si. However, new semiconductor materials and cell design techniques have now made it possible for a-Si thin-film PV cells to compete in conversion efficiency with other technologies at a much lower cost. Many believe that a-Si cells are the PV breakthrough we have been waiting for to make solar technology affordable. Efficiencies in the realm of 5% to 10%, and near-future costs in the neighborhood of $1.00 to $2.00 U.S. per peak watt, make this technology very attractive and practical. Compare this to peak watt costs of $3.00 U.S. to $10.00 U.S. for other technologies.

Figure 3-3 shows a continuous sheet module of thin-film a-Si PV cells, all manufactured together at the same time on the same substrate. Some manufacturers are able to produce continuous long strips of these cells on thin stainless steel sheets. That means these cells can be manufactured on flexible substrates that conform to shapes — a possible advantage for some applications. They can also be made as shingles for roofs of buildings and homes. Low cost and adaptability are definite strengths. One relative problem has been that of longevity. Useful life has not been as long as the typical 10 to 20 years of x-Si and P-Si cells. This will be corrected if it has not been done do already. Certainly companies such as Siemens Solar Industries of Camarillo, California, and Solarex of Frederick, Maryland are working on this and other problems to bring this technology to the market.

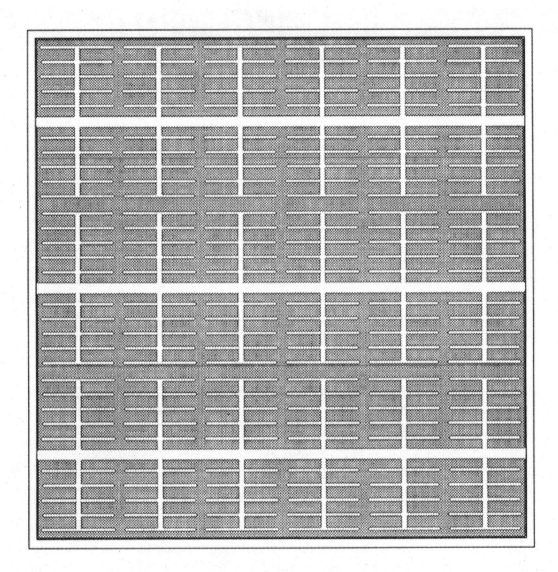

Figure 3-3. *A module of amorphous PV cells.*

PV Technology Projects and Programs

The U.S. Department of Energy (DOE) through one of its national laboratories, the National Renewable Energy Laboratory (NREL) of Golden, CO, under its Thin Film Partnership Program, has awarded millions of dollars to many U.S. companies and universities to move the very promising thin-film PV technologies forward. The following is a listing of awards made during 1995:

1. $3.5 million to Golden Photon, of Golden, CO, to improve manufacturing processes and efficiencies of thin-film a-Si photovoltaics.

2. $3.5 million to Solar Cells, Inc., of Toledo, Ohio to improve design and production of very promising cadmium-telluride (CdTe) thin-film technology PV cells.

3. $1.73 million to International Solar Electric Technology of Inglewood, California, to develop PVs based on copper indium diselenide (CuInSe or just CIS). They are trying to perfect a very inexpensive process for manufacturing acceptably-efficient CIS PVs.

4. $906,000 to the University of South Florida to develop processing/manufacturing methods for cadmium telluride (CdTe) and copper indium gallium diselenide (CuInGaSe) PVs.

5. $3.9 million, three-year contract, to Energy Photovoltaics, Inc., of Princeton, NJ to develop more efficient ways to coat glass with CIS thin films. (From *NREL In Review*, Summer 1995, pg. 8)

United Solar Systems Corporation of Troy, Michigan, was scheduled to complete one of the world's largest PV manufacturing facilities in Newport News, VA, late in 1995. With financial assistance from NREL/DOE, they have developed multilayered thin-film PVs on a "roll" (760 m. or 2500-foot stainless steel flexible sheet) that have a record energy conversion efficiency of 10.2%. They expect to be able to produce 10 megawatts of panels per year. United Solar manufactures both rigid and flexible thin-film panels. (From *NREL In Review*, Spring 1995, pg. 4.)

AstroPower, Inc., of Newark, DE, is another success story bred partly from DOE assistance. Over the past few years, they have set new standards for the power output and efficiency of single-crystal silicon (x-Si) PV cells and modules. They are well known for their large 6" square, high-output solar cells and high-wattage modules, and are among the fastest-growing PV manufacturers in the world. Take a few moments to look over the AstroPower data sheets and you will see what I mean. Also included are some Siemens data sheets, which we will discuss shortly:

AP-105 and AP-106 Solar Cells

AstroPower's AP-105 and AP-106 solar cells are high-efficiency five and six inch single crystal (Cz) solar cells. Single crystal solar cells have been more widely sold and utilized than any other type of solar cell in history. AstroPower Cz solar cells build on this impressive record of performance and reliability, but add unprecedented value by taking advantage of economies of scale. The AP-105 and AP-106 are the largest solar cells commercially available, and are fast becoming the solar cell component of choice for module manufacturers around the world.

Several technology improvements have been incorporated into these solar cells, including: the use of thicker wafers to improve yield through the module manufacturing process; burnished silver bus bars on both front and rear contacts for enhanced solderability; and a proprietary anti-reflection (AR) coating which provides superior color uniformity and enhances module appearance.

Whether you are looking to manufacture modules to meet the emerging industry standard power range of 65 – 75 watts, or wish to leap ahead of the pack into lower cost large area modules, AstroPower solar cells can help you meet your goal.

AstroPower AP-105 and AP-106 Solar Cells – higher power, lower cost.

Solar Cell Features

- Proven high efficiency single crystal (Cz) silicon technology.

- High power solar cells mean lower module cost – fewer solar cells to interconnect and package per peak watt of module output.

- Largest solar cells commercially available.

- Burnished silver bus bars for enhanced solderability.

- Unique anti-reflection (AR) coating provides extremely good color uniformity.

- More resistant to breakage due to use of thicker wafers.

AstroPower

AP-105

Rated Power	1.5 - 2.0 Watts
Open Circuit Voltage	0.53 - 0.58 Volts
Short Circuit Current	4.4 - 4.8 Amperes
Test Voltage	0.450 Volts
Current at Test Voltage	3.2 - 4.6 Amperes

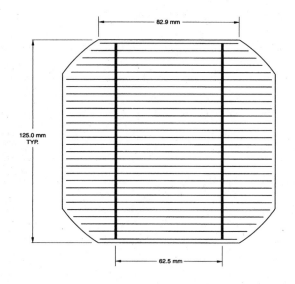

AP-106

Rated Power	2.3 - 3.3 Watts
Open Circuit Voltage	0.53 - 0.58 Volts
Short Circuit Current	7.1 - 7.8 Amperes
Test Voltage	0.450 Volts
Current at Test Voltage	5.2 - 7.4 Amperes

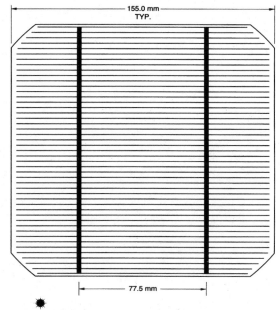

ASTROPOWER

ASTROPOWER, INC., SOLAR PARK, NEWARK, DE 19716-2000 USA TEL 302-366-0400 FAX 302-368-6474

AP-1106/AP-1206 Photovoltaic Modules

AstroPower's AP-1106/AP-1206 series modules utilize the new AP-106 high-efficiency six inch single crystal (Cz) solar cell, the largest solar cell commercially available today. Over the last two decades, Cz solar cells have become the industry standard –accounting for more than twice the installed capacity as any other PV technology. AstroPower's new six inch Cz solar cells extend this proven and reliable technology to a new realm of ultra-large area, capturing significant economies of manufacturing scale.

The high power of these modules makes them particularly well suited for use in large arrays, where mounting and interconnecting costs can be reduced compared to an equivalent capacity array employing a large number of lower power modules. This attribute is especially important for emerging utility applications. Still easily handled by a single person, these modules are also ideal for conventional industrial applications such as telecommunication.

The AP-1106/AP-1206 modules utilize industry standard construction techniques for high strength and durability. Every module is covered by a comprehensive 10 year warranty and meets all applicable industry and consumer standards for safety and reliability.

AstroPower AP-1106/AP-1206 modules.
The new standard.

Module Features

- Each module contains 36 series connected single crystal silicon solar cells for optimum battery charging performance in hot weather or low light levels.

- High power output per module means fewer modules to interconnect and mount.

- Over 6.5 amps of charging current in full sunlight.

- Heavy duty anodized frame provides strength and convenient mounting access.

- Module width and mounting hole pattern conform to industry standards – fits existing mounting racks and trackers.

- Weather resistant junction box, including protective diodes, allows for easy and safe field interconnection.

ASTROPOWER

Electrical Characteristics

AP-1106

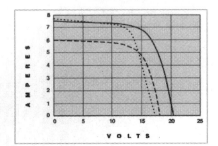

AP-1206

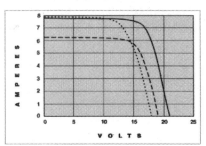

	AP-1106	AP-1206
Rated Power	110 Watts	120 Watts
Open Circuit Voltage	20.7 Volts	21.0 Volts
Short Circuit Current	7.5 Amperes	7.7 Amperes
Max. Power Voltage	16.7 Volts	16.9 Volts
Max. Power Current	6.6 Amperes	7.1 Amperes

—— 1000 W/m², 25 deg. T_{CELL}
······· 1000 W/m², 60 deg. T_{CELL}
----- 800 W/m², 45 deg. T_{CELL}

Mechanical Characteristics

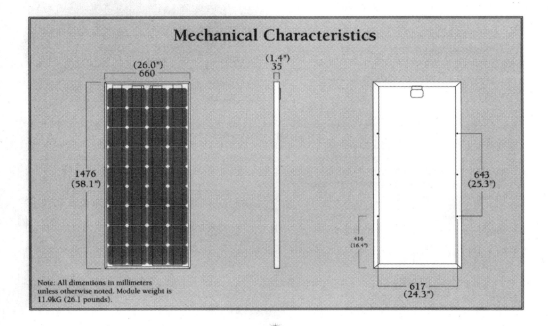

(26.0")
660

(1.4")
35

1476
(58.1")

643
(25.3")

416
(16.4")

617
(24.3")

Note: All dimentions in millimeters unless otherwise noted. Module weight is 11.9kG (26.1 pounds).

AstroPower

AstroPower, Inc., Solar Park, Newark, DE 19716-2000 USA tel 302-366-0400 fax 302-368-6474

SIEMENS

M65 Self regulating solar electric module

RATED POWER 43 WATTS

With 30 cells in series, the high efficiency Siemens M65 is a 43 watt, self regulating solar electric module. Self regulation eliminates the need for separate charge control devices, resulting in a simple, reliable and economical power generating system.

The M65 module regulates its electrical output to the needs of the battery. As the battery approaches full charge, it decreases its typical current charging rate of nearly 3 amps to less than a 1/2 amp.

Utilizing the highest standard of construction, the M65 module is able to withstand some of the harshest environments in the world and continue to perform efficiently.

Siemens solar electric modules are tested to meet or exceed industry standards, and even more rigorous Siemens quality and performance requirements.

10 YEAR WARRANTY

The Siemens M65 solar electric module carries a 10-year limited warranty on power output and is listed by Underwriters Laboratories (UL), an independent, not for profit organization, testing for public safety.

Siemens solar electric module features:

- Silent operation
- Sunlight as fuel
- High power density
- Easy installation
- Rugged, durable construction
- Simple, reliable operation
- Easy to expand systems
- Low maintenance
- No moving parts to wear out
- No environmental pollutants

M65 Self regulating solar electric module

FEATURES

Large, high efficiency single crystal solar cells provide the highest light to energy conversion efficiency available from Siemens.

Cells are textured and have an anti-reflection coating.

Multiple redundant contacts provide a high degree of fault tolerance and circuit reliability.

Cells within a module are electrically-matched for increased efficiency.

Circuit is laminated between layers of ethylene vinyl acetate (EVA) for moisture resistance, UV stability and electrical isolation.

Low iron tempered glass front for strength and superior light transmission.

Rugged anodized aluminum frame is designed for exceptional strength.

Side rails with multiple mounting holes for easy installation.

Tough, multi-layered polymer backsheet is used for environmental protection, resistance to abrasion, tears and punctures.

Two junction covers with lids are designed for easy field wiring, safety and environmental protection.

Wired-in bypass diodes reduce potential loss of power from partial array shading.

SPECIFICATIONS

Rated Power	43 Watts
Current (typical at load)	2.95 Amps
Voltage (typical at load)	14.6 Volts
Short Circuit Current (typical)	3.3 Amps
Open Circuit Voltage (typical)	18.0 Volts

Power specifications are at standard test conditions of: 1000 W/M² solar irradiance, 25°C cell temperature and solar spectral irradiance per ASTM E892

Weight	10.5 lb/4.8 kg

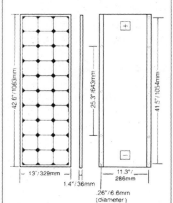

CHARACTERISTICS

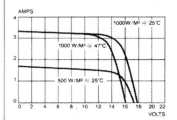

The IV curve (current vs. voltage) above demonstrates typical power response to various light levels at 25°C and a 47°C cell temperature.

- Minimum power upon final factory inspection is within 10% of rated power.
- Module leakage current of less than 50μA at 3000 VDC.
- Normal operating cell temperature (NOCT) as defined by ASTM E 1036 is 42°C +/- 2°C.
- Laboratory tested for wide range of operating conditions (−40°C to 90°C, 0 to 85% humidity).
- Passes Salt Fog Test per Mil-Standard 810.
- **Passes complete environmental requirements of JPL Specification No. 5101-161 (Block V).**
- External grounding screw for electrical safety.
- Ground continuity of less than 1 ohm for all metallic surfaces.
- Ten-year limited warranty on power output.*
- UL Listed. (Per UL 1703).

Charts are for estimating purposes only. Specifications subject to change without notice.

*Complete warranty and installation information is included in the module package or is available from Siemens or your Siemens Solar dealer prior to purchase.

Siemens Solar Industries
P.O. Box 6032, Camarillo, CA 93011
Telephone: (805) 482-6800   FAX: (805) 388-6395

©1995 Siemens Solar Industries. 111-700016-78 Rev. C (2173)   Printed in USA   6/95   Printed on Recycled Paper

In addition to being well-known for their high-performance, low-cost, x-Si PV cells, AstroPower is also recognized for their pioneering of a new, polycrystalline silicon PV cell and process. Named "Silicon-Film™," this technology involves a new proprietary process that enables polycrystalline silicon to be grown on inexpensive substrates using less silicon, and eliminating the waste and expense of boule growing and cutting, and wafer polishing. AstroPower's 6" square (240 cm^2) AP-225 cell is the pride of their accomplishments. This cell is capable of producing a little over 3 W of electricity at an efficiency greater than 12%, and they're not finished yet. Further improvements are continuing. A 20 kW test site containing 312 AstroPower APM-7225 modules, each composed of 36 early model AP-225 cells, was placed online at the PVUSA project in Davis, CA, on November 30, 1994. PVUSA is a joint photovoltaic test project funded by utility companies and the DOE and managed by Pacific Gas & Electric (PG&E).

Solarex Corporation of Frederick, MD is another exciting photovoltaic success story. Solarex is a business unit of Amoco/Enron Solar and is the largest U.S. owned manufacturer of photovoltaic cells and modules. Amoco/Enron Solar is a joint venture of Amoco Corporation (global producer and marketer of oil, natural gas, petroleum, and chemical products) and Enron Corporation (natural gas purchasing and marketing and electricity production). Solarex has been long known for their high-quality polycrystalline silicon PV cells, and are the world record holders for rooftop photovoltaic installations. The current record held by Solarex is a 500 kW rooftop installation on the Trade and Technology Center in Bonn, Germany. The second largest rooftop installation, also by Solarex, is the 349 kW system on the Aquatic Center (Natatorium) of Georgia Tech University, Atlanta, Georgia. This was the site for the swimming events of the 1996 Summer Olympics.

Solarex is also rapidly becoming known for its amorphous silicon, thin-film PV technology and manufacturing capability. In October of 1995, ground was broken for a new 83,000 square-foot thin-film PV manufacturing facility located in the Stonehouse Commerce Park of James City County, VA. Full-scale production of 10 million watts of PV modules per year is expected by mid-1997. Each a-Si, thin-film PV module will be approximately 8 square feet and produce about 56 W of electricity. With the addition of this facility, named TF1, Solarex expects to be the world's largest manufacturer of PV modules.

PV Arrays

As was mentioned earlier, conventional x-Si cells put out a peak voltage of nearly 0.5V in direct sun, the amount of current depending on size of the cell. To increase the voltage to the

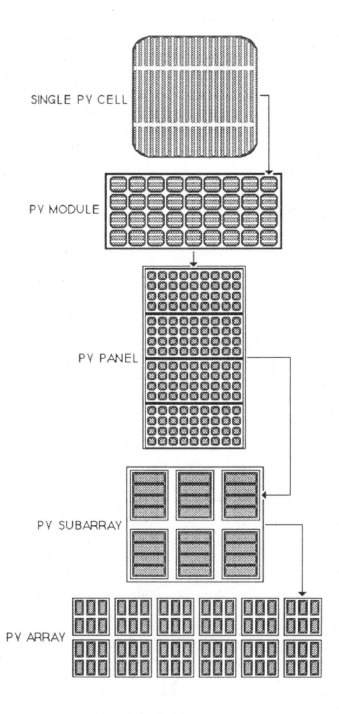

SINGLE PV CELL

PV MODULE

PV PANEL

PV SUBARRAY

PV ARRAY

Figure 3-4. *From PV cell to array.*

desired level, usually 14V to 18V per module, cells must be combined in a series (daisy-chain) arrangement, with the negative terminal of one cell connected to the positive terminal of the next cell, like "D" cells in a flashlight. (See the Siemens and AstroPower data sheets.) Then, these modules can be combined in a parallel wiring scheme to obtain a higher current forming what are known as *panels*. Actually, the modules in panels can be wired in series to obtain a still higher voltage if desired. The illustration in *Figure 3-4* further demonstrates how panels form subarrays, and combined subarrays form large arrays, sometimes referred to as *solar farms*.

PV Systems

Stand-Alone (SA) Systems

When an array is installed, with all of its wiring, associated electronics, and means for energy storage, a PV system is formed for practical use. *Figure 3-5* illustrates a typical PV system. The power delivered by the PV array is DC voltage and current. A voltage regulator is used to control the charge rate of a bank of storage batteries. The batteries are special heavy-duty deep-cycle (deep-discharge) types that provide for nighttime or cloudy-day operation. An inverter is used to convert the DC from the batteries to the proper AC voltage and frequency for appliances, lights, etc. A PV system such as this is often referred to as a "stand-alone" (SA) system because of the use of a frequency- and voltage-regulated inverter and storage

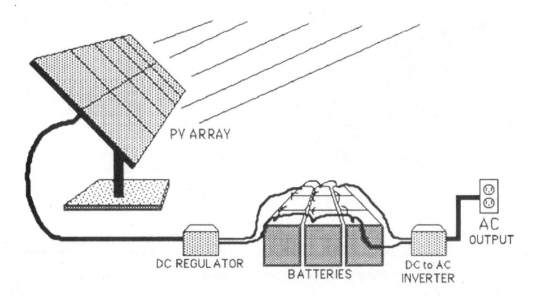

Figure 3-5. *A typical stand-alone battery-storage PV system.*

batteries. One of the greatest expenses involved in a PV system, besides the array itself, is the batteries. Some manufacturers of industrial-grade deep-cycle batteries boast of 10-year life-times. However, these batteries and individual cells are expensive. Their initial cost is high, maintenance is routine, and replacement is inevitable.

Utility-Interactive (UI) Systems

In some PV systems, batteries are not needed or desired. A home owner, for example, can actually use the local utility company for his PV energy storage instead of batteries, by using a *synchronous inverter*. The synchronous inverter is placed between the PV array and the utility company power meter. The inverter converts the DC from the PV array to AC, at the proper voltage and frequency, and backfeeds it through the power meter to the utility lines. In so doing, the inverter voltage is synchronized with the AC utility frequency, and the power meter is slowed down or stopped; or, in cases when the energy demand in the home is very low, the meter is actually reversed. (Certain types of watt-hour meters can be reversed). When the sun is shining brightly, energy flows back to the power company. At night or on cloudy days, energy is retrieved from the power company. Thus, the utility company pro-vides the energy storage and eliminates the need for battery storage. Systems that make use of the power grid for energy storage are often referred to as "utility-interactive" (UI). Today, many power companies actually pay customers a small amount for their excess energy.

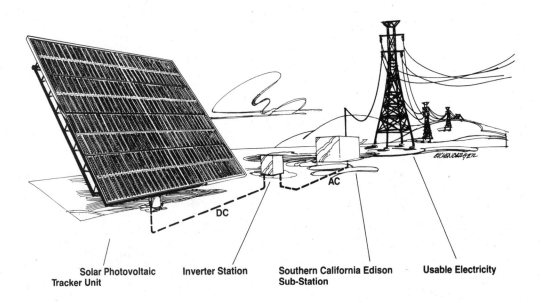

| Solar Photovoltaic | Inverter Station | Southern California Edison | Usable Electricity |
| Tracker Unit | | Sub-Station | |

Figure 3-6. A commercial solar PV installation. Courtesy of Siemens Solar Industries.

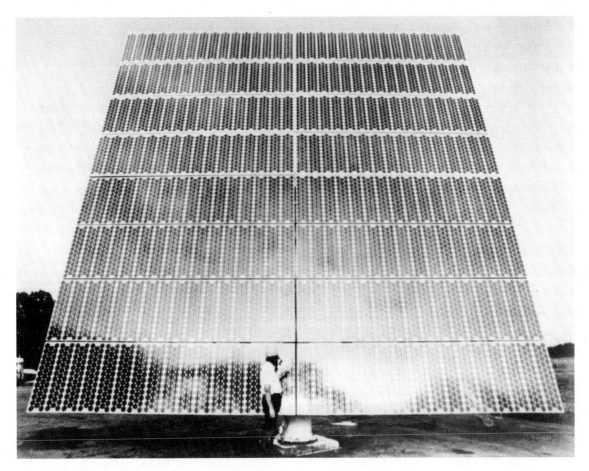

Figure 3-7. *One of the 108 tracking arrays at Hesperia, CA.*
Courtesy of Siemens Solar Industries.

The UI PV system previously described is also used on a very large scale. *Figure 3-6* shows a component sketch of a one-million watt PV installation at Hesperia, CA, originally built with PV modules from ARCO Solar, Inc., now owned by Siemens Solar Industries of Camarillo, CA. The sketch shows only one of the 108 tracker units that follow the sun from sunrise to sunset. The tracking technique increases overall system efficiency by about 40%. Each tracker unit holds 256 individual modules with a total peak power capacity of 9600 W. (See *Figure 3-7.*) The combined DC voltage and current from all the arrays is converted to AC, and synchronously fed into a 12kV utility grid. *Figure 3-8* shows an aerial view of the entire facility. Approximately 300 to 400 homes can be powered from this installation. Siemens Solar Industries is a leader in the development and implementation of solar technology. They are well-known for a wide variety of PV modules used as building blocks in installations throughout the world.

Figure 3-8. *The entire 1 MW PV facility at Hesperia, CA.*
Courtesy of Siemens Solar Industries.

Figure 3-9 shows a dramatic ground-level view of a 6 MW facility that was installed in the early 1980s at Carrisa Plain, CA. It was intended as a huge test facility and has since been dismantled. As in the lower power facility at Hesperia, CA, sun tracking arrays were used to increase overall efficiency. A further 50% increase in efficiency was obtained through the use of laminated glass reflectors skirting each panel. The entire facility was unmanned and totally computer controlled. Again, there was no means of energy storage since the DC was converted to AC and synchronously fed into the utility grid. One very interesting feature of these systems is that the tracking arrays were automatically placed in a horizontal position when the wind velocity exceeded 30 mph, in order to minimize wind loading. When the wind subsided, the arrays were returned to proper orientation with the sun.

Researchers have been busy devising new techniques to increase the conversion efficiency of silicon PV cells for many years. Sandia National Laboratories, in Albuquerque, NM, are

Figure 3-9. *6 MW facility at Carrisa Plain, CA. Courtesy of Siemens Solar Industries.*

Figure 3-10. *A Fresnel lens/PV array under test at Sandia National Laboratories. Courtesy of Sandia National Laboratories, Albuquerque, NM.*

known for their alternative energy programs. *Figure 3-10* shows a Sandia Laboratory scientist inspecting a small, liquid-cooled PV array in an early 1980s experiment. The array is composed of 12 miniature, 1-cm. diameter PV cells (not visible in the photo), each with its own 12.5 cm. x 12.5 cm. Fresnel lens to collect and focus the sun's energy. The overall peak

	1991	1995	2000	2010–2030
Electricity price (¢/kWh)	40–75	25–50	12–20	<6
Module efficiency* (%)	5–14	7–17	10–20	15–25
System cost ($/W)	10–20	7–15	3–7	1–1.50
System lifetime (years)	5–10	10–20	>20	>30
U.S. cumulative sales (MW)	75	175	400–600	>10,000

*Range of efficiencies for commercial flat-plate and concentrator technologies.

Photovoltaic progress and program goals (costs and prices in 1995 dollars)

Figure 3-11. *Photovoltaic program goals. Courtesy of the National Renewable Energy Laboratory, Golden, CO, and the U. S. Department of Energy.*

efficiency obtained was a respectable 17%. The cooling system is extremely important in a system such as this since the Fresnel lens magnifies the sun's rays tremendously.

PV Now and in the Future

With all of the talk of new, low-cost photovoltaic technologies, and the many companies manufacturing them, you can't help but ask, "Where are they? Why don't we see them in stores? Why can't we go to our local Radio Shack® and buy a solar panel system and install it on our home?" These are good questions. All of the PV technologies we have discussed are available now. There are many catalog-sales, energy-related, and retail outlets in existence right now. One very popular source for energy-related products is Alternative Energy Engineering of Redway, CA (800-777-6609). Hutton Communications, Inc., of Norcross, GA, is another source (800-741-3811). There are others, and many are showing up on the Internet; see the listing of web sites at the end of this chapter. However, the cost of these systems is still very high, which does not nurture a huge market of eager customers. Folks who do add solar PV to their homes do so at a premium and can afford it. The hope is that as manufacturing volume increases and processes continue to improve, the costs will eventually be attractive to a larger market. Then, you will see national chain stores begin to stock PV systems. Most residential PV installations are being handled by certain experienced contractors.

At present, and for the most part, manufacturers are selling their PV products to large utility companies (usually with government subsidies or other incentives), and to governments who have power needs in remote areas. The U.S. DOE has technologically and financially assisted many countries in the use of PV technology in an effort to open foreign markets for U.S. manufacturers. The greater the markets, the greater the production, and the lower the prices will fall. Many of these issues are discussed in a recent full-color booklet titled *PHOTOVOL-TAICS: The Power of Choice*, produced by the National Renewable Energy Laboratory for the DOE. It covers the national photovoltaics program plan for 1996 through 2000. Copies can be obtained from the National Technical Information Service, U.S. Department of Commerce, 5285 Port Royal Road, Springfield, VA 22161 (703-487-4650).

Figure 3-11 is taken from page 9 of the booklet, and summarizes where we've been, where we are, and where we hope to be with PV technology. Note that the cost per watt ($/W) is the overall system cost. As an example, the range of cost shown for 1995 is $7 to $15 U.S. This implies an average cost of $11. At that rate, a 5,000W household installation in 1995 would cost $55,000. If you are talking about a $250,000 (or greater) home, the additional cost is not that severe. However, strap that onto an $80,000 home and you have a significant additional cost. The hopeful projection is that by the year 2000, we will be able to buy the same system for $25,000 at an average cost per watt of $5. So, you see, for most of us, PV is still not affordable. The projected hope is that one day it will be. However, do not be surprised if the predictions in *Figure 3-11* experience further delay. In reality, it is not technology that is the problem: it is economics on a very large scale, which includes power companies, oil companies, and politics. In a later section of this chapter, you will meet people who are not waiting for all of this to get sorted out.

3-2. SOLAR THERMAL CONVERSION (STC) TECHNOLOGY

Solar thermal conversion (STC) employs the use of concentrating solar collectors that focus and convert the sun's energy to very high temperatures. The temperature at the focal point can range anywhere from 200 to 2000° F, depending on the system design. There are basically three types of collectors that are used in STC systems: parabolic troughs (*Figure 3-12*),

Figure 3-12. *An engineer at Sandia National Laboratories inspects a parabolic trough with a conversion efficiency of 70% and a 600° F focal-line temperature. Courtesy of Sandia National Laboratories, Albuquerque, NM.*

Figure 3-13. *Parabolic solar collectors focus the sun's rays to heat liquids used for thermal energy transfer. Courtesy of Sandia National Laboratories, Albuquerque, NM.*

parabolic dishes (*Figure 3-13*), and heliostat/central receiver systems (*Figure 3-14* and *Figure 3-15*). Sandia National Laboratories and the U. S. Department of Energy (DOE) have been at the forefront of this technology.

Distributed and Central Receiver Systems

There are basically two approaches to STC systems: the distributed collection system, and the central receiver system. The distributed collection systems make use of troughs and dishes to collect and convert the sun's energy. As in PV systems, the total amount of energy collected over an entire installation is distributed among individual solar tracking units. The distributed collection of energy is transferred to a power plant via water, oil, or liquid metal

Figure 3-14. At the National Solar Thermal Test Facility at Sandia National Laboratories, 222 heliostats focus solar energy onto the central receiver atop the 200' tower. The heliostats comprise 88,000 square feet of reflective surface (over 1.5 football fields). Courtesy of Sandia National Laboratories, Albuquerque, NM.

(sodium) in a massive plumbing network. In contrast to the distributed collection scheme, the central receiver system has only one point of energy concentration (the central receiver), on which many tracking reflectors (heliostats) focus the sun's energy.

Each of these two approaches has advantages and disadvantages. The biggest advantage to the distributed system is that it will collect usable quantities of energy from indirect light available on hazy or overcast days, whereas the central receiver system requires full sunlight. However, the central receiver system has two major advantages. First, light is used to transfer energy to the central receiver from a field of tracking reflectors. This eliminates the need for the massive, energy-wasting plumbing systems used by distributed systems. Secondly, the

Figure 3-15. *Here is a close view of one of the 222 heliostats of the National Solar Thermal Test Facility of Sandia National Laboratories. Courtesy of Sandia National Laboratories, Albuquerque, NM.*

temperatures attained in a central receiver system are much greater than that of the distributed collectors, 1000° F (540° C) or more as compared to 200° to 800° F (90° to 430° C) for distributed systems. The higher temperature allows steam turbines to operate at higher efficiencies in the generation of electricity.

Figure 3-14 and *Figure 3-15* show the 5 MW heliostat/central receiver system at the National Solar Thermal Test Facility of Sandia National Laboratories located on Kirtland Air Force Base in Albuquerque, NM. The heliostats can focus a peak flux of 260W/cm^2 and can illuminate target areas as large as 300 ft^2. This facility leads the way for the practical application of this technology.

The Solar One Project

To utilize, refine, and improve central-receiver technology, a 10 MW plant named Solar One was built at Barstow, CA, for use in assisting with peak load demand. It consisted of 1818 glass heliostats reflecting the sun's rays on a water-filled central receiver. Computers monitored the overall system to determine when the central receiver system should be brought on line. A conventional boiler, heated from fossil fuels was used to supply steam to the turbine generator, but the superheated steam from the central receiver could be used when sun was available and demand was high. This was a "next step" pilot project that ran from 1982 to 1988. It was a great success, and has led to further improvements. Wide-scale utilization by power companies is expected as we approach the year 2000 and beyond.

The Solar Two Project

Out of the Solar One project has come the final step before central receiver commercialization takes place. That step is the Solar Two project. This project involved the redesign and refitting of Solar One from a water/steam system to a molten salt system — a process that began when Solar One was shut down in 1988, and was completed in 1995. The same 300 ft. (91 m) central receiver tower and 1818 heliostats were used at the Barstow, CA, location.

The sun's energy, concentrated on the central receiver, heats a mixture of sodium and potassium nitrate, which melts at 430° F. The heated salt reaches temperatures over 1000° F (538° C) and circulates in a closed system. Part of the closed system is a huge, hot storage tank where the heat is either stored or converted to steam to drive a turbine generator. Cooled molten salt (550° F) is returned to the central receiver to gain heat energy and return to the 1000° F temperature range. This 10 MW facility is expected to be dwarfed by 100 MW and 200 MW plants by the year 2,000. However, experts within the DOE and utility companies believe that these systems must be hybrids combining solar and natural gas energy sources in

Figure 3-16. *The Dish/Stirling solar-thermal electric system is a joint venture of Sandia National Laboratory and Cummins Engine Company. Photo courtesy of Sandia National Laboratories.*

order to make them cost competitive. According to the National Renewable Energy Laboratory, in their Winter 1994/1995 *In Review* magazine, a 100 MW hybrid could produce electricity at a cost of 7 to 10 cents per kilowatt-hour compared to 12 to 16 cents for a solar-only system.

Dish-Stirling Solar Electric Systems

Another exciting and practical solar thermal project that has recently reached maturity is the Dish-Stirling system shown in *Figure 3-16*. This project has been a joint effort between Sandia National Laboratories and Cummins Engine Company. (Cummins is well known for making diesel engines.) The overall operation of the system is relatively simple, and boasts a solar energy conversion efficiency of 29%. As you can see in *Figure 3-16*, the dish is comprised of individual circular mirrors; each mirror made of a highly-reflective stretched membrane. The entire dish tracks the sun under computer control. At the focal point of the dish is a self-contained energy conversion system based on an advanced Stirling engine. This engine has only one moving part — a free piston that is gas lubricated, requiring no oil lubricants. Durability is extremely high. This piston is also the armature of a linear generator. The back-and-forth motion of the piston induces a nominal 240VAC, and more than 7 kW of power is generated! At the base of the engine assembly, and in the focal point of the sun's energy, is what is called a *heat pipe receiver*, and a chamber filled with sodium. The sodium is superheated and serves as a small heat reservoir to maintain operation during brief cloud interruption. The chain of energy transfer is from solar to hot sodium, to hot gas (helium, to drive and lubricate the piston), to mechanical, to electrical. When sufficient solar energy is not available, the engine can be heated by burning natural gas, heating oil, or propane. Under solar power, the system is totally clean and pollution-free. Cummins plans to market this new product for rural/remote applications.

3-3. SOLAR/OCEAN THERMAL TECHNOLOGY

Seventy percent of the earth's surface is covered by water, making the oceans the greatest solar collectors of all. Every year the oceans collect thousands of times more energy than man consumes. So, how can we recover this energy for our use? Basically, the answer lies with specially-designed, low-temperature, low-pressure heat engines. What is a heat engine? A heat engine is any device that develops mechanical energy from thermal energy. The overall efficiency of the heat engine is determined by the amount of thermal energy that is actually converted to mechanical energy, as compared to the total amount of thermal energy applied to the engine. Much of the thermal energy is expelled as waste due to design limitations. An automobile engine is example of a heat engine. Within the engine, chemical energy (gasoline) is converted to thermal energy (combustion), which yields mechanical energy and exhausted thermal energy. Jet engines, diesel engines, Stirling engines, and steam engines are all examples of heat engines; but these are all high-temperature, high-pressure engines, none of which are suitable for ocean thermal energy conversion (OTEC).

Ocean Thermal Energy Conversion (OTEC)

Most heat engines operate at high temperatures, creating rapidly expanding gases that, in turn, generate mechanical energy and motion. These high temperatures are not available in the ocean. Therefore, combustion or steam cannot be produced to drive a conventional heat engine. The surface temperature of tropical oceans is only about 84°F (29°C). However, this is warm enough to boil a liquid/gas, such as ammonia, freon, or propane, in a special chamber called an *evaporator*. The expanding, low-temperature gas is then used to power special low pressure turbines. The turbines then drive a generator that produces the electricity. The exhausted low-temperature gas is contained and condensed, using cold (roughly 35 to 40°F (1.5 to 4.5°C)) deep ocean water. A small pump recycles the liquid back to the evaporator

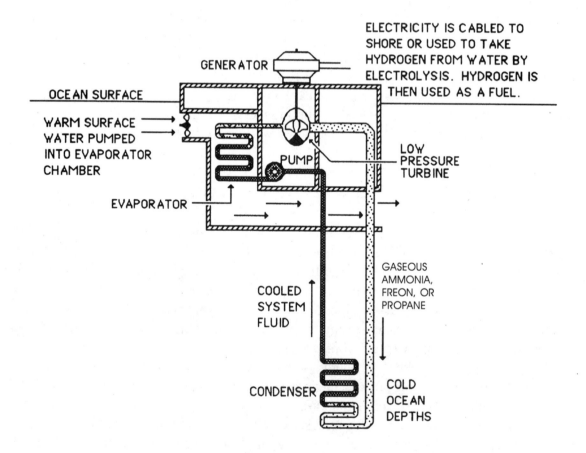

Figure 3-17. Basic OTEC system.

where it is once again converted to a gas. This closed-cycle is continuous, driven by the difference in temperature between surface and deep water.

Figure 3-17 is a simplified illustration of a floating platform OTEC system. In this illustration, the condenser is at a depth where the water temperature is very low. In most designs, the condenser is at the surface, and cold water is pumped from great depths. This may require pipes reaching to depths of 1000 ft. to 3,000 ft. (305 m. to 915 m.). The difference in gaseous pressure (created from the difference in temperature), between the evaporator and the condenser is the force that drives the turbine.

Surprisingly, small OTEC plants have been built at various times over the past 100 years. A French scientist by the name of D'Arsonval proposed the idea in 1881. Since then, scientists the world over have proposed and built OTEC installations of varying sizes, designs, and success rates. Some of the problems these researchers have had to contend with include: the high corrosiveness of sea water; growth of marine life that clogs pipes and heat exchangers; high winds and waves; great offshore distances between optimal generating sites and civilization; high initial costs; high maintenance costs; and low overall efficiency (about 2%). Though these problems persist, the idea of a free and inexhaustible supply of ocean thermal energy still challenges engineers and scientists today. Great progress has been being made toward high-output, low-maintenance OTEC systems.

The Natural Energy Laboratory of Hawaii Authority (NELHA)
73-4460 Queen Kaahumanu Hwy., Suite 101, Kailua-Kona, HI 96740
(808) 329-7341   Fax: (808) 326-3262

Initially established in 1974, the Natural Energy Laboratory of Hawaii Authority (NELHA) has played a vital role for the United States in OTEC research and development. NELHA is well established as a world leader in OTEC and related research. The laboratory is partly funded by the state of Hawaii and the U.S. Department of Energy.

The first successful floating OTEC plant was built and operated in 1979 through the efforts of NELHA and a private company. Dubbed Mini-OTEC, the plant was constructed on a converted Navy barge and tethered just off Keahole Point on Hawaii's west shore. While lasting only three months, the project clearly demonstrated that producing electricity from a 20°C difference in water temperature was indeed possible. It produced a total of 50 kW. Of that 50 kW, 33 kW to 40 kW was needed to run pumps and other essentials to operate the plant, leaving 10 kW to 17 kW as surplus. The project, though short-lived, also provided valuable insights into the seriousness of corrosion and the biofouling of systems.

A DOE-funded project was OTEC-1, a converted Navy tanker anchored off Kawaihae on the Kona Coast. This 1980 project was intended to test heat exchangers and the elements of closed-cycle OTEC systems. However, the project was not intended to generate electricity. Very little data was collected because the project was terminated after only four months of operation.

In 1983, a very large OTEC project was designed by a company called Ocean Thermal Corporation, under contract from the DOE. The plans were for a 40 MW OTEC plant to be constructed on an artificial island at Kahe Point off the coast of Oahu. $2 million was contributed toward the project from the State of Hawaii. Unfortunately, the Department of Energy cut off funding after the preliminary design phase and the project was halted. This was during a time when oil costs were low, and OTEC was not considered competitive with fossil fuels.

Since then, many experiments and developments have taken place at NELHA. Aluminum of Canada (ALCAN) and the Marconi Division of General Electric Company of Great Britain have been working to develop new aluminum condensers and evaporators for OTEC systems. Research into OTEC and many other projects continues at NELHA today.

210 kW OTEC Plant at NELHA

One of the greatest and most successful OTEC achievements to date is located, and fully operational, at the Natural Energy Laboratory of Hawaii, Keahole Point, near Kailua-Kona on Hawaii's west coast. It is a 210 kW test plant that has been in operation since 1992. It is the world's first continuously-operated OTEC plant. The plant is owned and operated by the Pacific International Center for High Technology Research (PICHTR) of Honolulu, and was jointly-funded by the state of Hawaii, PICHTR, and the U.S. Department of Energy. Its net (or usable) electrical production is in the neighborhood of 50 kW, a relatively modest amount. Approximately 160 kW of the total 210 kW is used by the process itself, operating pumps, etc. However, when scaled up to a plant size greater than about 5 MW, the net output will be more than 75% of the total gross output.

On the positive side, the 50 kW surplus of NELHA's OTEC plant is pollution-free electricity derived from an inexhaustible source. In addition, there are no harmful by-products resulting from the operation. On the contrary, a very useful by-product is pure, desalinated drinking water, thousands of gallons per day. In addition to the fresh water produced, the cold ocean waters brought from depths of over 2,000 ft. are rich in nutrients that are not found near the surface. This water is also free of diseases that are found near the surface. This cold and

Figure 3-18. *This is the 210 kW OTEC plant located at the Natural Energy Laboratory of Hawaii. Photo courtesy of the Natural Energy Laboratory of Hawaii Authority.*

nutrient-rich sea water is not only used to condense the steam in the OTEC plant, but used to cool the earth to grow non-native crops, to raise lobster, shrimp and other seafood, to grow algae and seaweed, and to air condition buildings. The spinoffs are wonderful and numerous.

The NELHA/PICHTR Keyhole Point OTEC project is what is called an *open-cycle system*. In this type system, warm surface water is pumped into a very low-pressure boiler. The low pressure allows the water to boil at normal room temperature. The resulting cool steam drives a low-pressure turbine generator. The exhausted steam coming from the turbine is cooled and condensed from cold deep-ocean waters, and is returned to the ocean or contained as pure drinking water. What happens to the salt? It's left behind in the vacuum evaporator chamber where the ocean water turns to steam. The salt does not accumulate in this chamber because it is continually flushed out back into the ocean waters. Less than 0.5% of the water pumped into the chamber is actually turned to steam.

The cold water for the OTEC system and other NELHA projects is pumped up through three different pipes from depths of 1913 ft. (583 m.), 2031 ft. (619 m.), and 2215 ft. (675 m.). From shore to the depths, the pipes measure around 6000 ft. (1830 m.) in length. Altogether, they deliver 17,100 gallons per minute to the facility. The water temperature arrives at the plant at nearly 43°F (6°C) after having increased in temperature by only about 0.5°C while being delivered. The warm water is piped in from near-surface depths of 50 to 70 ft. (15 to 21 m.). The temperature of the surface water is in the range of 77° to 83°F (25° to 28°C) all year. That provides an average difference in temperature of about 36°F (about 20°C), sufficient to allow the plant to operate effectively and reliably. All of these pipes tie into the plant sitting on the shore. Shown in *Figure 3-18*, the plant itself has the outer appearance of a concrete silo 31 ft. high and 22 ft. in diameter. The generator sits atop the silo structure.

PICHTR has recently finished another OTEC project at NELHA on Keyhole Point. This is a 50 kW closed-cycle plant funded by the state of Hawaii, PICHTR, the Hawaiian Electric Company, and the National Defense Center of Excellence for Research in Ocean Sciences at NELHA. The closed-cycle plant uses the aluminum heat exchanger recently developed by ALCAN (Aluminum of Canada). A large, 1 MW closed-cycle plant at NELHA has also been proposed by KAD Partners of Hawaii. The dream for the future is a Hawaii that is almost totally free of fossil fuel. Except for aircraft fuels, this is entirely possible in time since all electrical needs and land transportation (electric vehicles) can be powered using OTEC plants to harvest the sun's energy stored in the ocean.

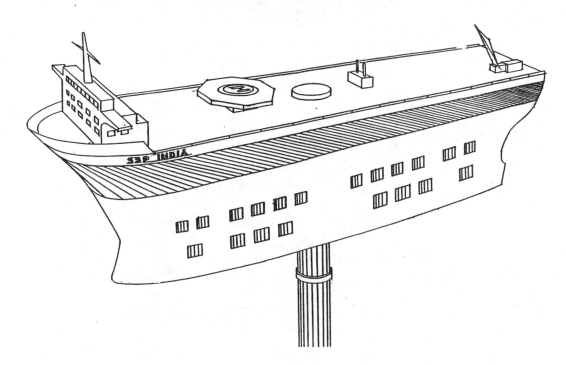

Figure 3-19. *Now under construction is a floating, 100 million Watt OTEC plant off the southern tip of India. Approximate hull dimensions: 150 m. long, 48 m. wide, and 45 m. deep below the surface. Courtesy of Sea Solar Power, Inc., of York, PA.*

OTEC at Kulasekarapattinam, Tamil Nadu, India

A project is now underway that will provide 100 million watts of clean electricity from ocean thermal energy to people in southern India. A memorandum of understanding was signed in 1994 between the Indian state of Tamil Nadu and Sea Solar Power, Inc., of York, PA. Sea Solar Power will install and operate the plant. Shown in *Figure 3-19*, the entire floating plant is similar in appearance to a huge oil tanker and will be positioned about 46 km. off shore from Kulasekarapattinam in the Gulf of Mannar, just off the southern tip of India, between India and Sri Lanka. The plant will be a closed-cycle OTEC system.

Figure 3-20 is a simple schematic of the closed-cycle OTEC system designed by Sea Solar Power, Inc. Warm (above 80° F) surface water is pumped in through an above- and below-surface filter network, and is then piped down to a low-temperature boiler (evaporator). The evaporator contains propylene (a low-cost, noncorrosive refrigerant), which evaporates, cre-

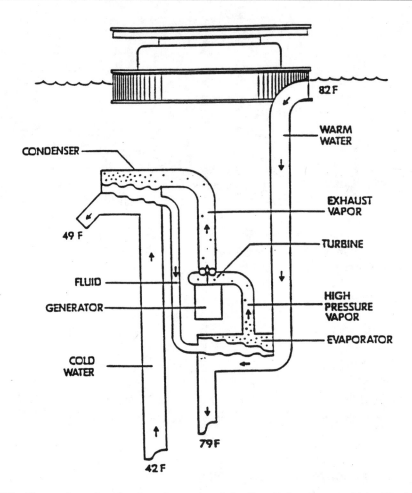

Figure 3-20. *Shown here is a basic schematic of the Sea Solar Power 100 MW OTEC system. Courtesy of Sea Solar Power, Inc., of York, PA*

ating pressure to drive a turbine generator. The entire plant contains twelve such turbine generators. The generators turn at 1500 rpm., producing 50 Hz AC and about 8.5 MW. Exhausted vaporous propylene continues upward to a condenser, where it is cooled and returned to the liquid state. This cooling and condensing is accomplished by the cold (42°F) water that is pumped from the ocean depths. The liquid propylene then returns to the boiler (evaporator), where it reheats and repeats the cycle.

Some Interesting Facts About the Kulasekarapattinam Project

The pipe shown in the center of the plant in *Figure 3-19* is used to draw cold ocean water from depths of 1,000 meters (about 3300 ft.). Its inner diameter will be almost 8 m., and it

Figure 3-21. *The Brevard Community College Solar Cruiser Project Team. Left to right, top row: Scott Jorden, Luke Halleman, Jim Watson, Tom Chaffee. Bottom row: Rob Waterman, Kevin Glaser (project manager), Ed Bollenback, Bill Sikinger, Terry Dye, George Weaver, Mark Hazen (Professor), and Bob Stein (not in photo). Photo courtesy of Brevard Community College, Palm Bay, FL.*

will be lowered through the hull in sections. Each section will be made of fiberglass, and the overall pipe will be flexible in nature. To aid in stopping corrosion, most of the equipment in the hull will be surrounded with fresh water. Scuba divers will perform routine maintenance inside the hull. Since most of the plant will be under the water line, it will be very stable in rough seas. Anywhere from 12 tons to 20,000 tons of fresh water will be produced per day during operation. This is because warm sea water that is taken in at the surface will first be deoxygenated to reduce corrosion and prevent the growth of sea life in the system. In this process, water vapor will be produced that is condensed and can be stored or transported to the mainland. In addition to the production of fresh water as a desirable by-product, Sea Solar Power, Inc., has predicted that a huge fishing industry will sprout up around the plant because of fertilizers pumped up from the depths that sustain an upward food chain. They believe that income from fish could be as high as $70 million per year. Underwater cables will carry 69,000 V from the plant to the shore. The entire plant will remain fixed in position using satellite information and rudders that direct exhausted water from the boilers and condensers.

3-4. SOLAR ENERGY USERS' SHOWCASE

The Solar Cruiser Project

This solar project is of special interest to me since it was designed and built by my students in their Project Management and Engineering Class at Brevard Community College in Palm Bay, FL. The Solar Cruiser is a 20-foot long aluminum pontoon boat with a roof of photovol-

Figure 3-22. Shown here are the roof-mounted solar panels of the Brevard Community College Solar Cruiser. Photo courtesy of Brevard Community College, Palm Bay, FL.

taic cells, four storage batteries, and two 40-pound-thrust electric trolling motors. *Figure 3-21* shows the proud crew of designers and builders, while *Figure 3-22* shows a bird's-eye view of the roof-mounted panels. Mr. Kevin Glaser was the project manager, and did an excellent job of keeping the team focused and on time. Starting from a salvaged aluminum pontoon platform, the students designed, built, modified, and fully documented the project in a total of only 16 weeks (Jan. 4 to May 10, 1994)! Great work, class!

An interesting feature of the craft is the means of control. All control functions of steering, variable speeds forward, and variable speeds reverse, are accomplished with a common computer-type joystick. The cruiser can be controlled effortlessly from any place on its deck. Handling the cruiser is an easy task even for one person.

The purpose of the Solar Cruiser Project is to provide a meaningful and in-depth design experience for graduating students from BCC's ASEET program. Course title: *Project Management and Engineering* EETC-2930

Specifications:

1. Means of propulsion: Two mechanically-connected (tie-bar linked) Motorguide 40 lb., 12V trolling motors. Each motor demands 30A.

2. Power control: Pulse width modulation, 20 kHz switching rate, power MOSFET "H" configuration, computer joystick used for all functions, full range of variable speeds (forward and reverse), control from any position on the craft.

3. Steering: Linear DC servo motor, 6-inch travel, screwshaft drive.

4. Energy storage: Four Stowaway 800 batteries from GNB.

5. Energy source: Roof-mounted PV array; 24 panels arranged to provide over 15V and a total maximum current of 50A

6. Cruise range: 5 hours on batteries only at full speed; 10 hours and more with solar panels, on a sunny day — no worries.

Mr. Solar's Solar-Powered Homestead

How many of us have said, "I'd sure like to live totally independent of all utility companies, and not have to pay for water, sewer, and electricity!"? Well, that might be nice if your well

Figure 3-23. *Mr. Solar: Charlie Collins and his wife, Fran, enjoy the independence that solar power provides. Photo courtesy of Charlie and Fran Collins.*

water is good, your septic system is well-designed, and you've taken the time to plan your energy requirements carefully. Let me introduce you to a clever couple who has done just that: Charlie and Fran Collins of La Verkin, UT. There they are in *Figure 3-23*, proudly poised in front of the fruits of their own labor. Charlie is well-known around the world as Mr. Solar, and their homestead, on the top of Smith's Mesa, is known as the Do It Homestead. ("If it needs done, you do it!")

The Do It Homestead began back in 1974 when Charlie hand-picked a 240-acre site on the top of Smith's Mesa with a breathtaking view of Zion National Park. (Background of *Figure 3-23*.) Later, Charlie met Fran and they were married in 1976. Together, they combined their many skills to make a home of virtual independence. Between the two of them, nothing is impossible. To say both are handy is an understatement. They enjoy hard work—everything from candle making, to growing their own food, to grinding cereal, to making furniture, to

Figure 3-24. *Fran Collins shows off the Collins' Siemens solar PV array.*
Photo courtesy of Charlie and Fran Collins.

making cloth from the wool of their sheep, to installing windmills, and installing modern photovoltaics. The list could go on! If you are connected to the Internet, you can visit Charlie and Fran at their homestead on the World-Wide Web. (http://www.netins.net/show-case/solarcatalog/)

The Collins' Energy System

Charlie and Fran have three different sources of electrical energy: photovoltaic array, windmill, and a backup generator. The PV array is their primary source of energy. As Fran is pointing out in *Figure 3-24*, the array consists of 24 Siemens M55 modules. Each module

Figure 3-25. *In preparation for the 1996 Summer Olympic Games, workers are installing PV panels on the roof of the Aquatic Center (Natatorium) of Georgia Tech University, Atlanta, Georgia. Photo courtesy of the National Renewable Energy Laboratory, Golden, CO, Craig Miller Productions, and the U.S. Department of Energy.*

produces approximately 17.5V at around 3A of current. The modules are interconnected in such a way as to provide nearly 35V and about 40A. The entire array is mounted on a dual-axis tracking system that adjusts azimuth and elevation to keep the array directly facing the sun. This additional tracking system is well worth the extra initial cost since it gains more electricity than it uses. The PV electricity travels through heavy-gauge cables to a regulator/charger, then on to the storage cells. The regulator/charger is necessary to control the voltage and current applied to Charlie and Fran's huge bank of storage cells, which need only about 27V to 29V during charge. There are 40 nickel-iron storage cells in the bank providing 600 ampere-hours of capacity—enough energy storage for three or four days of cloudy-day usage. The DC energy from the storage bank is supplied to a 4,000W inverter that converts 24VDC to 120VAC. Charlie and Fran have many common household appliances that utilize

Figure 3-26. *Solarex Technical Center, Frederick, MD: a photovoltaic-powered manufacturing facility. Electricity from its 200 kilowatt array provides uninterruptible power to certain critical manufacturing processes. Photo courtesy of Solarex, Frederick, MD.*

the 120VAC, such as a dishwasher, trash compactor, vacuum cleaner, etc. In addition to the PV array, they have a 12V windmill system and a backup generator. Charlie reports having to use only 12 gallons of fuel in his backup generator for a year.

Solar at the 1996 Summer Olympics in Atlanta, Georgia

Figure 3-25 shows workers installing PV panels on the roof of the Natatorium, the Aquatic Center at the University of Georgia, used for the 1996 Summer Olympics. As of this writing, this is the world's second largest roof-mounted PV array, totaling 349,000 watts on the building and entrance. The main roof array is 340,000 watts consisting of 2,856 PV modules, each producing a peak power of nearly 120 watts. This array covers 3,680 square meters. Using

Figure 3-27. *South View of the Fiore Home in Englewood, FL. Note the PV array and solar water heating system on the roof. The PV array consists of 6 large panels. The four large panels near the roof peak are solar water heating panels. To the far left are solar water heating mats used to heat the pool. Photo courtesy of Tony and Nancy Fiore.*

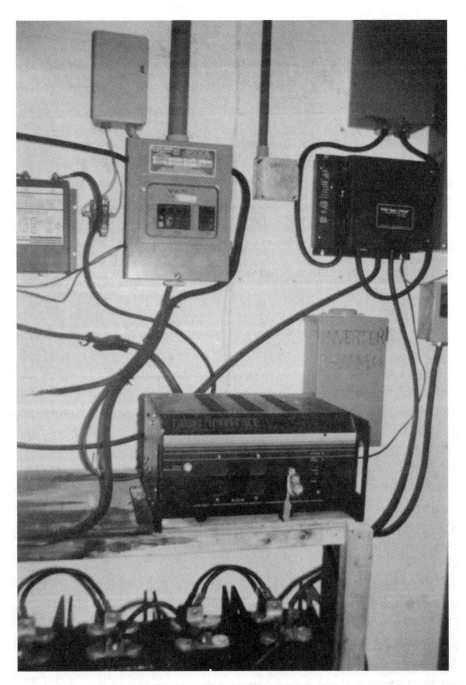

Figure 3-28. *Here we see Tony and Nancy's storage bank (partially shown at bottom), 3000W inverter (sitting on bench), regulator/charger (upper right square box on wall), and various circuit breakers and disconnects. Photo courtesy of Tony and Nancy Fiore.*

an insolation factor of 1000 W/m², the total available solar power over the 3,680 m² is 3.68 million watts (3,680 m² x 1000W/m² = 3.68 MW). From this calculation, we can determine the overall energy conversion efficiency to be a little over 9% (340 kW/3.68 MW = 0.092). The actual efficiency of each PV cell is much higher, since the overall area of just the cells is less than the area of the entire array, which includes gaps and hardware.

This, and the world's largest roof-mount record, is held by Solarex of Frederick, MD. Solarex is well-known for their high-efficiency polycrystalline silicon (P-Si) PV cells. Shown in *Figure 3-26* is the Solarex Technology Center, their international headquarters and P-Si PV manufacturing facility. Their roof PV array produces a peak output of 200,000 watts and is used to power certain critical manufacturing processes. Solarex has other manufacturing facilities in the U.S. and Australia, and offices around the world.

Tony & Nancy Fiore of Englewood, Florida

Tony and Nancy Fiore have lived on and have enjoyed solar energy for over 10 years in their Englewood, FL, home. They own Solaray Systems, a photovoltaic engineering company, and Fiore Construction, both of Englewood, FL. Their home is definitely unique in their residential area. Using proven energy conservation techniques, Tony and Nancy designed and constructed their home to be very comfortable and virtually free of the need for air conditioning. Then, they carefully designed their photovoltaic system to meet the energy needs of their family of four.

Figure 3-27 shows the rear view of Tony and Nancy's home. Their roof-mounted PV array is rated at about 1400 peak watts. Heavy cabling brings this energy down to a regulator/converter which controls the charge to their bank of storage cells. The large storage cells, partially shown at the bottom of *Figure 3-28*, are arranged in a daisy-chain configuration to provide a nominal 24V with enough storage for several days of cloudy weather.

The 24V bank is connected to a 3,000W DC-to-AC inverter which provides their home with 120VAC. The inverter, shown in *Figure 3-28* on the wooden bench just above some of the storage cells, is a high-efficiency modified sine wave DC-to-AC converter. Also, part of their overall system is a 24VDC-to-12VDC converter that supplies 12V of electricity to some lighting and a refrigerator. Finally, the least popular addition to the system is a 12V generator bicycle that sits in their living room. When ridden, it feeds 12VDC back into the storage bank. Needless to say, they do not depend on this!

Figure 3-29. *Three teams participated in the trial run for Sunrayce '95. The finish line of the trial and actual race is the Solar Energy Research Facility of the National Renewable Energy Laboratory (shown in the background) in Golden, CO. Photo courtesy of the National Renewable Energy Laboratory, Golden, CO, Warren Gretz, and the U.S. Department of Energy.*

Sunrayce: Solar Car Race

Sunrayce is a biennial event sponsored by the U.S. Department of Energy (DOE) and managed by one of its national laboratories. The event is co-sponsored by many corporations, such as General Motors (who began the event), Electronic Data Systems Corporation, Midwest Research Institute, the Environmental Protection Agency (EPA), the Society of Automotive Engineers, Sandia National Laboratories, and the National Renewable Energy Laboratory. General Motors sponsored the first Sunrayce, called *Sunrayce USA*, in July, 1990.

The first DOE-organized event took place in 1993. The emphasis of Sunrayce is on education, energy and the environment, and is open to college-level institutions of North America.

Sunrayce '95 was the second biennial DOE-sponsored race. The race began on June 20, 1995, in Indianapolis, IN, and ended on June 29 at the National Renewable Energy Laboratory in Golden, CO. First place went to the Massachusetts Institute of Technology, whose total elapsed time was 33:37:11, with an average speed of 37.23 mph. Second was the University of Minnesota at 33:56:00 and 36.88 mph. Third place went to California Polytechnic University, Pomona, with an elapsed time of 37:03:43 and an average speed of 33.77 mph. In addition to the top three trophy positions, the DOE presented scholastic achievement awards for technical innovation, engineering excellence, artistic talent, teamwork, and good sportsmanship.

Some 65 or more teams registered for the race, making it the largest such event to date. Each had to prepare and submit a formal proposal to the DOE that detailed their solar car project. A team of experts (formed from the sponsors) evaluated the proposals and selected 30 team entrants as "seeded" teams. These seeded teams each received $2000 from DOE and $1000 from the EPA to assist them with initial expenses. Prior to the June 20 race date, the teams had to qualify for the 1100 mile cross-country event. From the qualifying race, 40 were selected to compete in the main event. Those who did not qualify for the main event were invited to participate in exhibition races in some cities along the race route. Racing took place from 10 AM until about 7 PM each day.

Figure 3-29 gives you an idea of what most of the vehicles looked like. The vehicles are entirely solar-powered. They gather their energy from the sun and store it in on-board batteries. Locomotion is accomplished with a 2- to 10-hp electric motor. Great care and consideration must be given to every element of the design to ensure low friction, low rolling resistance, low wind resistance, and efficient energy utilization. During the race, great skill is needed to utilize collected and stored energy wisely. Battery capacity was limited to 5 kWh. Each vehicle also included safety features such as a horn, tail lights, a rear vision system, and seat belts. Many other electrical, mechanical, and procedural regulations are specified for each biennial event. All the details of Sunrayce '95 are available on the Internet at http://www.nrel.gov/sunrayce. Details regarding Sunrayce '97 can be accessed at this site as well.

3-5. REVIEW QUESTIONS

1. Explain how energy released from the sun is transported to the earth.
2. List three sources of energy that can actually be considered forms of solar energy.

3. What is the approximate amount of solar energy available per square meter on the surface of the earth assuming a clear day?
4. A PV system for a home in a remote location where there are no utility lines will use _____ for energy storage.
5. What is the approximate increase in overall PV system efficiency when a solar tracking system is used?
6. What is a central receiver solar thermal conversion system?
7. List three examples of heat engines.
8. What does OTEC stand for?

Check your answers in the *Answers to Questions* appendix at the back of this book.

3-6. WORLD-WIDE WEB SITES

The Author's Home Page

http://www.castlegate.net/personals/hazen

Relating to Photovoltaic Energy Conversion

AAA Solar Service and Supply
    http://www.rt66.com/aaasolar/

Alternative Energy Engineering, Inc.
    http://www.alt-energy.com/aee

American Solar Energy Society
    http://www.csn.net/solar/

Centre for Photovoltaic Devices and Systems
    http://www.vast.unsw.edu.au/pv.html

El Paso Solar Energy Association Home Page
    http://www.realtime.net/~gnudd/react/epsea.htm

International Solar Energy Society
    http://www.ises.org/

Mr. Solar's Home Page
    http://www.netins.net/showcase/solarcatalog/

Maine Solar House
    http://solstice.crest.org/renewables/wlord/index.html

National Renewable Energy Laboratory
    http://www.nrel.gov/

N.C. Solar Center
    http://www.ncsc.ncsu.edu

NW-SolarCell
    http://www.seanet.com/Users/miknel/NW-SolarCell.html

Photovoltaic Technology
    http://www.sandia.gov/Renewable_Energy/photovoltaic/pv.html

Solar Electric House, The
    http://www.ultranet.com/~sda/bookinfo.html

Solar Energy
    http://www.energy.ca.gov/energy/earthtext/solar.html

Solar Energy - Raymond J. Bahm
    http://www.rt66.com/rbahm/

Solar Energy Network, The
    http://204.214.164.14/indexa.html

Solarex of Frederick, MD
    http://www.solarex.com

Solstice (Solar in Indonesia)
    http://solstice.crest.org/renewables/indonesia/index.html

Sunny Gleason and Solar Powered Systems
    http://www.cove.com/~sunny/

Sunrayce 95
    http://www.nrel.gov/sunrayce/

World Solar Challenge
    http://www.engin.umich.edu/solarcar/WSC.html

Relating to Solar Thermal Energy Conversion

Sandia National Laboratory
    http://www.sandia.gov/

Relating to Solar/Ocean Thermal Energy Conversion

Energy from the Ocean
    http://zebu.uoregon.edu/ph162/l17small.html

Natural Energy Laboratory of Hawaii
    http://bigisland.com/nelha/

4

HARNESSING THE WIND

Mankind has utilized the wind as a source of energy for thousands of years. The earliest boats used crude sails to capture the wind as first fishermen then explorers made their way across lakes and oceans. Early inventors discovered the wind could be used to pump water and grind grain into flour. One of the earliest references to windmill power is found in the history of the Persians during the 7th century. From there, wind-powered machines spread throughout the Middle East and Far East, finally appearing in Europe sometime during the 11th century. By the 1600s, the Netherlands had become the leading industrialized country in the entire world due to wind power. The Dutch were very clever by using wind to not only grind grain into flour, but to power factories and pump water. The Dutch people are renowned for their use of dikes and windmills to claim vast areas of land from the ocean.

In the 1800s, great growth of the wind power industry was seen in the United States and the rest of the world. The utilization of wind energy reached a peak by 1900, and began a period of decline into the first half of the 20th century. The steam engine and the internal combustion engine gradually replaced the wind machines of the world. The power-on-demand capability of these engines, combined with relatively low-cost fossil fuels, seemed to close the door on a wonderful era of wind machines. The early 1900s was also a period of growing dependence on electricity. Homes and factories were provided with electricity supplied by coal and hydroelectric plants. Small-scale wind-powered electrical generating systems were being used in rural areas on farms until rural electrification became practical. In a few locations throughout the U.S., Europe and Russia, large-scale wind-powered generating systems were being used to provide electricity to small towns. These systems were far from perfect and had to compete with the more reliable steam-powered plants that used economical fossil fuels. As a general rule, the governments of the world did not give serious consideration to wind-generated electricity until the oil embargoes and subsequent fuel shortages of the early 1970s. The embargoes provided tremendous incentives for many governments to earmark millions of dollars for the development of wind-generated electricity.

Research continues today as both government and private industries work to improve and further utilize this ancient source of energy, to help satisfy modern insatiable demands for electricity. As of 1995, it has been estimated that there are over 17,000 commercial wind turbines in the United States alone. Of those, 15,000 are in California, in three primary locations:

1. Altamont Pass, east of San Francisco
2. Tehachapi, northeast of Los Angeles
3. San Gorgonio, east of Los Angeles.

According to the Wind Energy Commission of California, these sites now produce about 1.1% of the total electricity needs of the state, and about 54% of the wind-generated electric-

Figure 4-1. *Actual operating experience demonstrates the successful use of intermittent renewable energy technologies. The Pacific Gas and Electricity system in California operates normally when wind generation supplies up to 8% of its system demand. Photo courtesy of the National Renewable Energy Laboratory, Golden, CO, Warren Gretz, and the U.S. Department of Energy.*

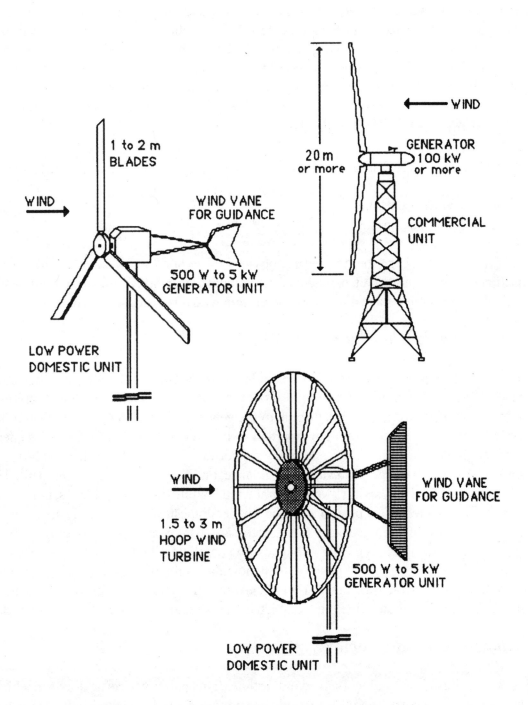

Figure 4-2. *Horizontal-axis wind turbines (HAWTs).*

ity in the entire world. *Figure 4-1* shows a very large wind farm owned and operated by Pacific Gas and Electricity in Altamont Pass, CA. Predictions by the U.S. Department of Energy (DOE) indicate that wind-generated electricity in the U.S. will increase six times in the next fifteen years. The American Wind Energy Association (AWEA) predicts that at least 10% of the total U.S. electricity needs will be wind-generated by the year 2050.

4-1. COMMON WIND MACHINES

Wind machines can easily be separated into two broad categories based on general design:

1. Horizontal-axis machines.
2. Vertical-axis machines.

The distinction between the two is very obvious. The horizontal-axis machines are those machines whose wind-driven mechanisms are mounted on a horizontal drive shaft, while the vertical-axis machines have their mechanisms mounted on vertical drive shafts.

Horizontal-Axis Wing Turbine (HAWT)

Figure 4-2 illustrates a few of the more popular horizontal-axis machines. In these designs, the generator unit is mounted on top of a tall mast or tower. Gearing is generally used to increase shaft rpm from the wind turbine to the generator itself. A special swivel mount is needed to allow the generator unit to turn freely with the wind, and permit electrical energy to be transferred to the wiring in the stationary supporting structure. HAWTs must be forced to face the wind using a large wind vane for guidance; or, in the case of many commercial units, they can be directed downwind by the natural wind loading of the huge blades, or they can be electrically or hydraulically positioned. The smaller domestic units generally use a wind vane mounted on a lockable hinged arm. If it is necessary to take the unit out of service, the hinge is unlocked and the vane is swung into a position parallel to the turbine blades. This forces the turbine to turn to a neutral position perpendicular to the wind (not facing). As a result, the turbine will stop rotating. In the commercial units, the blades are controlled hydraulically and can be "feathered out" (the pitch of each blade is set at 90°) to relieve rotational torque.

Vertical-Axis Wind Turbine (VAWT)

Figure 4-3 illustrates a vertical-axis wind turbine. This particular vertical-axis machine is a cylindrical sail, and works on the principle of a common wind-speed anemometer. One vertical half of the cylinder is concave, and the other half is convex. Wind pressure is greater on

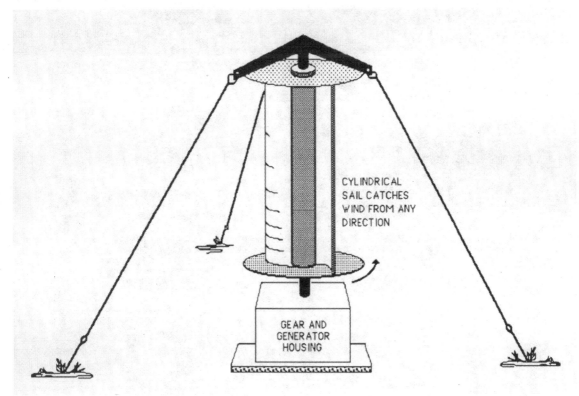

Figure 4-3. *Vertical-axis cylindrical sail system.*

the concave half, thus creating rotational torque. While the designs have only two vertical sections, some cylindrical sails have three. As you might suspect, VAWTs respond to wind from any direction. This may seem an advantage over horizontal-axis designs. However, in practice, this design presents a real problem in terms of braking or halting turbine rotation in any significant wind. In large systems, some means must be provided to equalize wind pressure on the cylinder or block the wind altogether.

Generally speaking, VAWT designs have several advantages over HAWT systems. One of the more significant advantages is lower cost. VAWT systems do not require tall supporting structures, which significantly reduces cost. A second advantage is related to the first: VAWT systems are simpler in overall design, which again holds cost down. In many HAWT systems, hydraulic or electrical control is needed to maneuver the turbine directly into the wind, and a blade pitch control system is needed to control speed in high winds. A third significant advantage is the placement of all mechanical and electrical apparatus at ground level. This allows for easy preventative maintenance or repair.

Figure 4-4. *Darrieus wind turbine located at the Sandia National Laboratories in Albuquerque, NM. Courtesy of Sandia National Laboratories.*

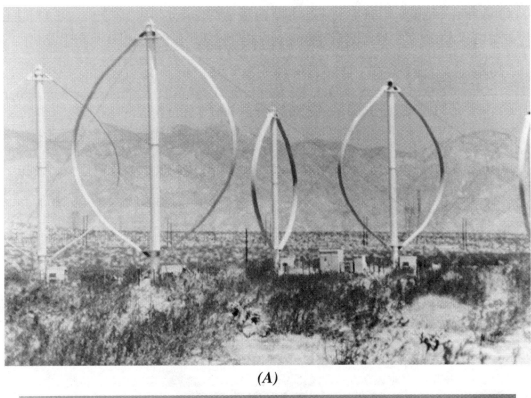

(A)

(B)

Figure 4-5. *Shown here is a grouping of vertical-axis wind turbines (VAWTs, photo A) and many of the thousands of horizontal-axis wind turbines (HAWTs, photo B) in the Cochella Valley near North Palm Springs, California.*

One of the more popular VAWT designs is shown in *Figure 4-4*. This strange-looking machine is known as a *Darrieus wind turbine*. The curved blades are aerodynamically designed (airfoil shaped, like an airplane wing) to create a difference in blade pressure, which causes rotational torque. The Darrieus turbine in *Figure 4-4* is 17 m. tall and produces a full 60 kW of electrical power in a 32 mph wind (51.5 km/h). One peculiarity of the Darrieus design is the requirement for a starter motor. An auxiliary motor at the base of the turbine is used to start the blades rotating. A small anemometer at the top of the structure provides a computer control system with wind speed information. If the wind speed is below 10 mph (16.1 km/h), this particular Darrieus cannot be started. If the wind speed is above 60 mph (96.6 km/h), the Darrieus is taken out of service (parked).

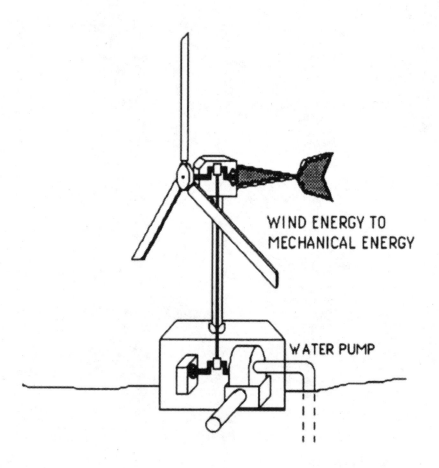

Figure 4-6. Wind energy to mechanical energy.

A. Stand-Alone System

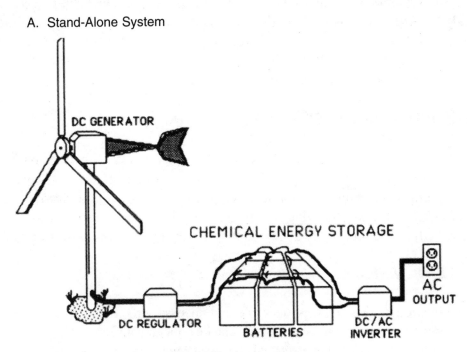

B. Utility-Integrated System (Utility Interactive)

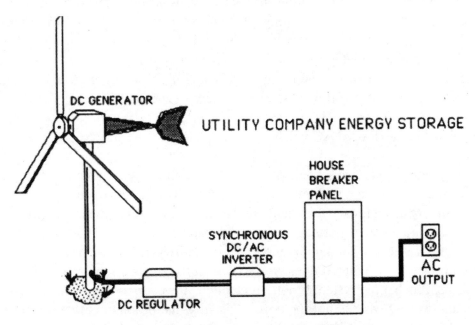

Figure 4-7. *Wind energy to electrical energy.*

Figure 4-5 shows some of the thousands of wind turbines that have been placed in the always-windy Cochella Valley near North Palm Springs, California. The valley is quite literally filled with thousands of HAWTs and hundreds of VAWTs for as far as you can see, on either side of Interstate Highway 10. This is a fertile proving ground for such systems since a strong wind blows continually through the valley.

4-2. WIND-POWERED SYSTEMS

Many wind turbines used in rural areas accomplish a specific task without converting wind energy to electricity. As shown in *Figure 4-6*, wind energy is converted to mechanical energy and used to pump water. In some applications, it may be desirable to accomplish a task directly instead of converting the wind energy to mechanical then to electrical. Such applications would be those that definitely do not require electricity or a steady flow rate of energy.

Chemical Energy Storage

In many applications, however, it is desirable and necessary to convert the wind energy to electrical energy. In doing so, the wind energy can be collected and stored as another form of energy and converted as needed to electrical energy. *Figure 4-7* illustrates two very common methods of collecting and storing wind energy for use as needed. The first method is *chemical energy storage*, *Figure 4-7a*, creating what is known as a *stand-alone (SA) system*. In a chemical storage system, wind energy is converted first to mechanical energy to drive a generator. The generator output is regulated and used to charge a bank of batteries. The electrical energy is converted to chemical energy and heat as the batteries are charged. The batteries will then provide a steady flow of electrical energy as needed. In many cases, the battery DC is converted to 50 or 60 Hz AC using an inverter. The inverter will convert the DC battery voltage to the desired AC voltage and frequency.

The chemical energy storage system does work well but has some fairly serious problems. One of the more significant problems is system cost. The batteries are very expensive and must be routinely cared for and eventually replaced (every 5 to 10 years). Another disadvantage is that of limited energy storage. When the batteries are fully charged, any additional wind energy is wasted. You might argue, the wind is free anyway, so why care? The wind does not always blow. When it does, you will want to collect all the energy you can. The wind may be free, but the system is not. In practice, however, the problem of fully-charged batteries and wasted wind is not the norm. Usually, the problem is that of undercharged batteries. Either way you look at it, fully charged or not charged enough, there is a problem that needs a solution. One such solution is the use of a synchronous inverter.

Utility Company Energy Storage

The battery (chemical) energy storage problems previously discussed can be solved by using the local utility company as a means of electrical energy storage. As shown in *Figure 4-7b*, this can be done by using a synchronous inverter to convert the wind-generated DC to AC at the proper voltage and frequency, synchronized with the power line. This method is often referred to as *utility-interactive* or *utility-integrated (UI)*. The wind system AC from the synchronous inverter is connected in parallel with the utility power at the house breaker panel. While the wind is blowing, all or part of the house energy consumption is provided by the wind. The demand for electrical power from the utility company, through the watt-hour meter, is reduced or eliminated. If there is more wind-generated electrical power than what is being used in the home, the watt-hour meter may, in some cases, actually be driven backward. In this way, the watt-hour meter reading may be reduced, indicating less consumed power from the utility company. This is a practical form of energy storage in which the utility company uses your excess wind-generated electrical energy and gives you credit for it. Thus, the costly batteries are eliminated, along with the space, weight, and maintenance problems.

4-3. TECHNICAL PROBLEMS & SOLUTIONS

Wind Availability

Wind availability is perhaps the most obvious and significant problem with wind systems. In most areas of the United States, "energy winds" only blow about 2 days out of a week. "Energy winds" are winds having speeds of 10 to 25 mph (16 to 40 km/h). Most of the week, the prevailing winds range below 10 mph. Commercial units require energy winds for moderate outputs. Recall the Darrieus VAWT at Sandia Laboratories that requires a 32 mph wind for maximum output (10 mph minimum for any output). Most wind systems operate only about 25% of the time at 50% or less capacity. This naturally makes the cost per kilowatt-hour fairly high. The solution to this problem is to place wind turbines in regions that have a history of sustained energy winds and to use wind energy in conjunction with other sources of energy.

Wind Constancy

Wind constancy is another serious problem facing wind energy utilization. The wind does not blow at a constant speed throughout the day. The wind is extremely unpredictable and variable. Rapid variations in wind speed place tremendous stresses on the wind turbine and cause it to vary its speed radically. This makes it difficult to regulate voltage and frequency in systems that use AC generators (alternators). The solution to the problem of wind constancy is manifold and costly. Mechanical mechanisms called *constant speed drives*, or *converters*, are employed to help regulate alternator rotation speed under varying turbine speed conditions. Also, instead of a single, extremely-high power wind turbine, a network of relatively low-power wind turbines can be erected over a large open plain. All units can be interconnected and computer-controlled in order to help overcome wind variations.

Energy Storage

Energy storage is another problem that has been discussed already. When the wind is blowing, the energy must be captured and either used immediately or stored for later use. Since energy storage is a problem common to many forms of alternative energy sources, we will address this topic in a separate chapter later.

4-4. AMERICAN WIND ENERGY ASSOCIATION (AWEA)

The following is based on information supplied by the American Wind Energy Association, Washington, DC. You can visit them on the Web at http://www.econet.org/awea/.

American Wind Energy Association 122 C Street, NW, 4th Floor Washington, DC 20001.
Phone: (202) 383-2500
Fax: (202) 383-2505
E-mail: windmail@mcimail.com

Who and What the AWEA Represents

AWEA is a trade and advocacy organization based in Washington, D.C. It represents the U.S. wind energy industry and individuals who support clean energy in the legislative/regulatory, public relations, and international arenas. Its annual Windpower meeting is the only U.S. national conference devoted solely to wind energy, and features technical papers from experts in all areas of the field. AWEA offers business and individual memberships.

Since 1974, AWEA has worked to further the development of wind energy as a clean and reliable energy alternative. The association believes wind energy is a technology that is economically and technically viable and will play a major role in providing clean energy for growing needs throughout the world.

To serve its members most effectively, AWEA focuses its efforts and resources on five strategic objectives, including:

1. Policy development and legislative advocacy.
2. Communications and public relations.
3. Overcoming market barriers.
4. Technology development.
5. Education.

Association activities which support these objectives include:

1. Legislative and regulatory representation.
2. A pro-active communications program.
3. A wind energy export program.
4. An industry standards program.
5. A research and development committee.
6. A weekly newsletter for business members. (*Wind Energy Weekly*)
7. A monthly newsletter for all members. (*Windletter*)
8. An annual industry conference.
9. Regional seminars for groups such as utilities, environmental groups, and small wind system users.

Interesting Facts from the AWEA
(From *The U.S. Wind Energy Industry*, AWEA, Feb. 1995)

The wind energy industry has grown steadily over the past 10 years, with American companies competing aggressively in energy markets across the nation and around the world. This growth has been the result of a partnership with the U.S. Department of Energy (DOE) and its national laboratories: National Renewable Energy Laboratory in Golden, CO, and Sandia National Laboratories in Albuquerque, NM. As a result, a full range of highly-reliable and efficient wind turbines have been developed that boast of 98% availability in the field. Because of this phenomenal growth and technological achievement, wind power developers are now bidding utility-scale projects at prices as low as 3 to 5 cents per kilowatt-hour levelized over the life of the project. Compare that to the 30 cents per kilowatt-hour of installations in 1981. Utilities around the country are recognizing the benefits of adding wind generation to their systems and many realize that wind energy will provide one of the cheapest sources of power within the next decade.

From the beginning of extensive use of wind power in the U.S. around 1980, wind generation production has increased steadily. In 1994, a little less than 3.5 billion kilowatt-hours of wind-generated electricity was produced, and in 1995, about 3.53 billion kilowatt-hours. The interest in wind power among utilities and individuals continues to grow. Utilities across the country are turning to wind as a means of diversifying their energy resource base, balancing fuel supply and cost risks, and protecting community environments at a low cost. The United States has, by far, the largest installed wind energy capacity. Numbers from 1994 show that the U.S. capacity was 1,717 MW, followed by Germany at 643 MW; then Denmark at 540 MW; India at 182 MW; the Netherlands at 153 MW; and the United Kingdom at 147 MW. Many other countries are utilizing wind energy as well, with totals less than 100 MW.

As utilities and their customers enjoy the benefits wind brings to their energy supply, they and nearby communities are benefiting from economic development generated by the wind energy industry. About $3.5 billion ($3,500,000,000) is invested in the U.S. industry, where watt-for-watt, dollar-for-dollar, the investment creates more jobs than any other utility-scale energy source. In 1994, wind turbine and component manufacturers contributed directly to the economies of 44 states, creating thousands of jobs for American communities.

Now, here is a special news release from AWEA that should be of great interest to us all:

Earth Day 1996: An AWEA News Release (April 19, 1996)
(From Wind Energy Weekly #681, April 23, 1996)

In recognition of the 26th annual celebration of Earth Day, the American Wind Energy Association (AWEA) today released the following statement:

As our minds turn again to the increasing impact of human activities on the natural environment which sustains us all, it is important to note the continuing evidence of environmental damage from one of the most pervasive of those activities, the combustion of fossil fuels for energy and electricity supply.

Many experts and policy makers are coming to realize that wind and other renewable energy sources and energy efficiency offer increasingly economical, clean and sustainable alternatives to fossil technologies.

Fossil fuels are harmful in many ways:

1. Air pollution from particulates, produced by the burning of fossil fuels and other materials, causes thousands of deaths and costs the American economy some $11 billion each year, according to a report released in mid-1995 by the American Lung Association. The report, "Dollars and Cents: The Economic and Health Benefits of Potential Particulate Matter Reductions in the United States," says that nearly 2,000 premature deaths nation-wide would be prevented each year if particulate air pollution were reduced to levels con-sidered legally acceptable in California.

2. Roughly half of the 2,000 to 3,400 tons of mercury emitted worldwide each year from human activities comes from fossil fuel combustion. Many states have found it necessary to issue warnings and guidelines to limit the eating of fish from fresh water lakes due to the accumulation of mercury. Mercury is highly toxic — an amount as small as half a pound distributed among one million fish is enough to pose a danger to pregnant women.

3. New findings point once more to acid rain, which is caused by atmospheric emissions of sulfur dioxide and nitrogen oxides from power plants and autos, as being extremely harm-ful to forest ecosystems. The most recent data, from researchers at Syracuse University and the Institute of Ecosystem Studies at Millbrook, NY, is summarized in the "Science Times" section of the April 16 issue of The New York Times.

4. Climate change, believed to be the result of rising atmospheric concentrations of car-bon dioxide emissions from fossil fuels (as well as other "greenhouse gases"), is impli-cated in regional changes that are also harmful to trees. Environmental policy analyst and journalist Charles Little writes in his recent book, The Dying of the Trees, that a warming trend in Alaska has coincided with the decline of cedars there, while rising sea levels in Florida are killing sabal palms along a 200-mile stretch of the Gulf Coast. (An article by Little from Earth Island Journal that summarizes his book is available on the World-Wide Web at <http: //www.earthisland.org/ei/ journal/americas.html>.) Climate change is also, of course, a potential threat to many other ecosystems and a multitude of endangered species. The Intergovernmental Panel on Climate Change (IPCC), an international body that has studied the question extensively, maintains an excellent resource with many fact sheets and other materials on the Web at <http:// www.unep.ch/iucc.html>.

While the Clinton Administration has laudably developed a voluntary Climate Change Ac-tion Plan (CCAP) to encourage U.S. industry to reduce carbon dioxide emissions, its own forecasts from the Energy Information Administration's (EIA) Annual Energy Outlook, 1996, predict that CO_2 production from electricity generation will rise by nearly one-third over the next 20 years, as aging nuclear units are retired and replaced by new fossil-fired power plants. (The EIA forecast is available on the Web at <http://www.eia.doe.gov/oiaf/aeo96/ ele_envi.html>.)

To avoid the additional environmental damage that continuing acceleration of fossil fuel use implies, we must press ahead with the development and use of renewable energy and energy efficiency technologies.

A single 500-kilowatt utility-scale wind turbine can displace the emission of more than 800 tons of carbon dioxide each year (based on the U.S. average power plant fuel mix) along with dozens of pounds of acid rain-producing chemicals and particulates. The very modest amount of wind energy installed in the United States today displaces approximately 2.6 million tons of CO_2 emissions annually.

Wind energy has dropped in cost by over 80% since the early 1980s, and is today in a competitive range with other electricity generation options. As a result of continued progress, wind and other renewables are beginning to be viewed as serious long-term energy supply options.

5.   The Royal Dutch/Shell Group, in a widely cited future energy scenario (one of two developed last fall that look ahead to the year 2050), sees the possibility of a global renewable energy industry doing $150 billion in business annually 30 to 40 years from now, with solar energy eventually becoming the fastest growing energy source. (A summary of the Shell forecast is available through AWEA's Web site at <http://www.econet.org/awea/wew/ othersources/ otheroutlook.html>.)

6.   The IPCC has developed a set of future options for reducing global carbon dioxide emissions. Because of a near certainty that energy demand will rise over the next several decades, all options (even those that foresee continued heavy use of fossil fuels or nuclear power) require rapid growth in renewable energy deployment. By the year 2025, in the nuclear forecast (which is LEAST favorable to renewables), wind and solar would generate about 3.2 trillion kWh around the world annually, or more electric power than the U.S. uses today. By the year 2100, they would supply 9.5 trillion kWh.

7.   The Utility Wind Interest Group (UWIG), an organization of U.S. utilities who support this promising new energy option, says, "With today's wind turbine technology, wind power could supply 20% of this nation's electricity," adding that, "0.6% of the land of the lower 48 states would have to be developed with wind power plants. This area, about 18,000 square miles, is about the size of four counties in Montana. Furthermore, less than 5% of this land would be occupied by wind turbines, electrical equipment, and access roads. Most existing land use, such as farming and ranching, would remain as it is now."

We call upon U.S. legislators, regulators, and policymakers at the state and federal levels to take prompt and decisive action to begin the transition away from fossil fuels, and toward a sustainable energy future that is compatible with the natural environment.

4-5. WIND MACHINES AND MANUFACTURERS

Note: Addresses and other means of contact for the following manufacturers are provided in an appendix at the end of this book.

Utility-Scale Machines, 100 kW or More

Advanced Wind Turbines, Seattle, WA
(Characterization: Wind turbine manufacturer, large systems.)

Advanced Wind Turbines has built a strong reputation on their AWT-26 wind turbine shown in *Figure 4-8*. This turbine has a two-blade rotor with a diameter of 26.2 meters (86 ft.). The maximum output power for the AWT-26 is 275 kW. Unlike some systems that must be mechanically steered into the wind, the dual-blade rotor of the AWT-26 spins downwind of the tower. As a result, the overall design is less complex. However, the blades of the AWT-26 must operate in the wind shadow of the tower. This is usually a serious problem with two-blade machines since the shadow causes repeated impulse stressing (wind shear) of the blades as they move into and out of the tower's shadow. However, this shadow problem is over-come in the AWT-26 systems by aerodynamic treatment of the tower and a unique rotor mechanism that allows it to teeter on a pin and damper system similar to that used on helicop-ter rotors. The blades of the AWT-26 are fixed and can not be feathered for speed control as in some designs. Again, this simplifies the design. Made of wood and epoxy, the blades can flex to accommodate variations in wind speed. To prevent overspeeding, a small vane is mounted on the tip of each blade. The vanes are activated by the control system and serve as an air brake. Advanced Wind Turbines is also marketing a slightly larger, yet similar, system, the AWT-27. The AWT-27 has a 27.4 m diameter rotor and is designed to cut in at a slightly lower wind speed. Advanced Wind Turbines is a turbine manufacturing subsidiary of FloWind Corporation of San Rafael, CA.

FloWind Corporation, San Rafael, CA
(Characterization: Wind turbine manufacturer, large systems, developers, operators.)

FloWind Corporation has been involved in the sales and installation of many large wind tur-bine farms in California, selling such turbines as the AWT-26 previously discussed. However, they are well known for their own unique vertical-axis wind turbines (VAWTs) of the Darrieus type. A field of FloWind advanced VAWTs is shown in *Figure 4-9*. These have a higher height to diameter ratio than previous designs and are referred to by FloWind as EHD VAWTs

Figure 4-8. *This is the AWT-26 high-power wind turbine manufactured by Advanced Wind Turbines. Notice the AWT-26 operates downwind while other wind turbines (shown lower in the photograph) must be steered into the wind. Photo courtesy of Advanced Wind Turbines, Seattle, WA.*

(extended height to diameter ratio). FloWind is the only manufacturer of this type VAWT. As mention in an earlier section, these VAWTs have the advantages of immediate response to wind from any direction, heavy components on the ground, and ease of maintenance. A possible disadvantage is that winds are not usually as strong near the ground. However, this is minimized by proper placement of the VAWT.

Figure 4-9. *Here we see a field of advanced vertical-axis wind turbines designed by FloWind Corporation of San Rafael, CA, with assistance from the U.S. Department of Energy. Photo courtesy of the U.S. Department of Energy, National Renewable Energy Laboratory, and FloWind Corporation.*

Kenetech Windpower, San Francisco, CA
(Characterization: Wind turbine manufacturer, large systems.)

Kenetech Windpower has been hailed as having the most efficient wind turbine on the market. Their KVS-33 has a 33-meter (108 ft.) diameter rotor with variable-pitch blades. It is rated between 300 and 405 kW. Unlike many other designs, the KVS-33 is variable-speed (hence the VS in KVS). So, how does it maintain constant frequency and voltage? It doesn't and wasn't intended to. As the blades accelerate with increased wind speed, generated AC frequency and power also increase. Twin generators produce a variable-frequency AC that is rectified to DC then inverted to 60 Hz AC using an electronic inverter as discussed previously in this and other chapters. The result is a high-quality 60 Hz AC synchronized to the power grid. The KVS-33 operates over a 9 to 65 mph wind range.

Northern Power Systems, Moretown, VT: A New World Power Technology Company
(Characterization: Wind turbine manufacturer, large systems.)

Northern Power Systems of New World Power Technology Company has developed a two-blade wind turbine with a motorized mechanism for pointing the rotor into the wind (facing the wind). The North Wind 250 (for 250 kW), shown in *Figure 4-10*, eliminates any effect from wind shear caused by tower shadowing since the wind passes through the rotor first, in front of the tower.

The result is smoother, less stressed, and quieter operation. Similar to the AWT-26 of Advanced Wind Turbines, the North Wind 250 has two tiltable blades mounted on elastomeric teeter bearings. These blades also have small ailerons that are used to adjust speed and quickly respond to gusts of wind and allowing a more precise control of rotor speed. The 250 starts producing electricity in as little as 9 mph winds and reaches its rated output of 250 kW at a wind speed of 29 mph. The operational life of a North Wind 250 is rated at 30 years.

Zond Systems, Tahachapi, CA
(Characterization: Repair and maintenance, control systems, monitoring devices, rotors/blades, wind measuring instruments, gearboxes, nacelles, towers appraisal, construction, design, engineering, feasibility, insurance, marketing, meteorology, regulatory, repair, technical consulting, testing, wind resource assessment developers, operators.)

Zond Systems is well known for manufacturing the world's largest wind turbines, the Z-40 and Z-46. The number following the Z indicates the diameter in meters of the turbine's three-blade rotor configuration (Z-40 = 40 meters = 131 ft., and Z-46 = 46 meters = 151 ft.). To give you an idea of the size of these turbines, the Z-40's total weight is about 55,000 lbs., and its housing (containing drivetrain, gearbox, and generator) is larger than a family van. Each Z-40 is capable of producing 500,000W of electricity, enough to power over 100 homes (at 5,000W or less per home). While the Z-40 is well-tested, proven, and heavily deployed, the Z-46 is new and is in its testing phase now. Zond turbines are constant-speed turbines that produce a steady 60 Hz AC in varying winds. Speed is adjusted by changing the angle of the blades. The rotor can be brought to a halt by feathering the blades parallel to the direction of wind flow. This is done automatically in winds exceeding 65 mph. Each blade also has an aileron, much like that on the wing of an airplane, to assist in speed control in the presence of gusting wind.

Small-Scale Machines, Less Than 100 kW

Machines suitable for farms, homes, cabins, boats, etc.

Bergey Windpower Company, Inc., Norman, OK
(Characterization: Wind turbine manufacturer, small systems.)

Bergey Windpower Company was formed in 1979 and set out to produce the highest quality wind turbines in the world. They now claim to be the world's leading supplier of small wind turbines up to 10 kW in capacity. Their wind turbines are now operating in all 50 states and 70 countries around the world. Bergey Windpower is proud of their record of success fostered by strong customer support and highly-reliable wind turbines boasting of 4 to 6 years of operation requiring no maintenance.

Bergey wind turbines are available in three sizes—850 W, 1500 W, and 10 kW. All of their turbines have few moving parts. There are no gears, pulleys, belts, or chains. The generator itself is of the permanent-magnet armature type with stationary field coils and no brushes or slip rings to wear out. The turbine's three blades are made of a very flexible fiberglass, capable of many years of service.

Like other small-capacity turbines, Bergey wind turbines are suitable for a wide range of applications such as:

1. Remote homes & facilities.
2. Village electrification.

Figure 4-10. *The North Wind 250 shown here is a new advanced design incorporating many innovations. Photo courtesy of New World Power Technology Company/Northern Power Systems, Moretown, VT.*

3. Drinking-water pumping.
4. Stripper well pumping.
5. Hybrid systems (with PV, etc.).
6. Telecommunications.
7. Cathodic protection.
8. Irrigation systems.
9. Distributed generation systems.
10. Utility bill reduction.

Southwest Windpower, Flagstaff, AZ
(Characterization: Wind turbine manufacturer, small systems.)

Southwest Windpower has been producing battery-charging wind generators for over ten years and just recently introduced the AIR wind module. It is a breakthrough in micro wind turbine technology. Most turbines require expensive towers, external regulators, maintenance, and special equipment to install them. The AIR requires no maintenance, no heavy tower, is internally regulated, and can be installed in a little over an hour with no special tools. During its development, several breakthroughs were made. These include:

1. *BLADES.* The AIR's blades are made of carbon fiber reinforced composite that twists as the turbine reaches its rated output. This twisting changes the shape of the blade causing it to go into the stall mode. This limits the RPM of the alternator preventing damage in high winds. Additionally, the blade tip is less than 3 mm. thick, which allows for almost silent operation.

2. *ALTERNATOR.* The AIR's alternator is optimized to match as close as possible to the energy available in the wind. It is constructed of neodymium-iron-boron permanent magnets and is brushless for superior performance and maintenance-free operation.

3. *REGULATION & CONTROL ELECTRONICS.* The control electronics maintains a load on the alternator at all times to make sure that the turbine never overspeeds regardless of whether the battery is charging or not. As the battery charges, the regulator periodically checks the line correcting for voltage losses and monitoring charge rate. Once the battery is fully charged, the alternator is switched to a load.

The AIR marine is shown in *Figure 4-11*. It starts producing power in as little as a 6 knot wind and delivers 300 watts of power in 24 knot winds. The AIR marine has fully sealed

Figure 4-11. *The Air Marine wind turbine is shown here mounted shipboard on a simple mast structure. Courtesy of Southwest Windpower, Flagstaff, AZ.*

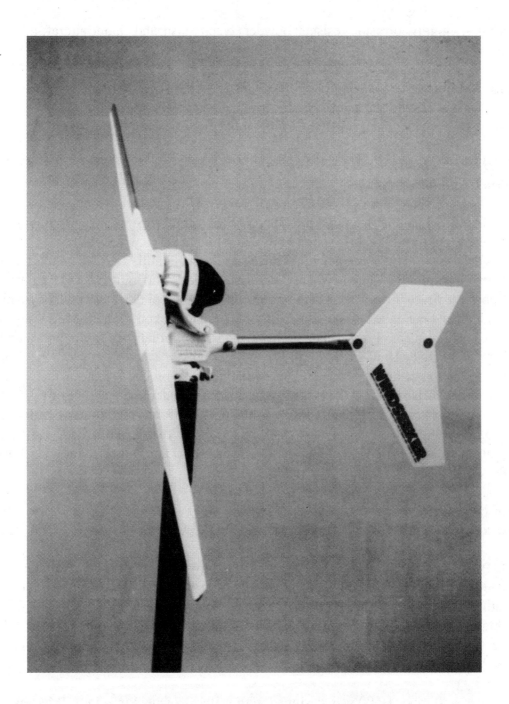

Figure 4-12. *The Windseeker wind turbine is suitable for home and cabin use. Courtesy of Southwest Windpower, Flagstaff, AZ.*

electronics, a highly corrosion-resistant powder coated finish, and unique aircraft-quality cast aluminum construction.

Southwest offers other models of wind turbines as well. Their series of WINDSEEKERS have been popular sellers and are available in a wide range of voltages (12, 24, 32, 48, 60, 90, and 120VDC). One of their WINDSEEKER models is shown in *Figure 4-12*.

WINDTech International, L.L.C., Atlanta, GA
(Characterization: Windmill manufacturer, water pumping, high volume.)

WINDTech International manufactures the OASIS-3 and OASIS-7 water-pumping wind-mills. These are incredible low-speed mills that pump large volumes of water in low winds from as deep as 4000 ft. These machines have a traditional multi-vane design reminiscent of open-prairie windmills. That is where their strength lies. The multi-vane design produces great power in low winds. The OASIS windmills begin pumping water in winds as low as 5 mph (8 kph) and continue pumping in as little as 2 mph (3.2 kph). The OASIS design is the crowning achievement of more than 10 years and $10 million of research conducted in the United States and the United Kingdom. It combines proven and verifiable oil-field counter-balance technology with rugged and reliable windmill technology to create an incredibly effi-cient and inexpensive new power source. To get an idea of their efficiency, one of WINDTech's OASIS-7 models can pump over 1000 U.S. gallons (3785 liters) of water per day from a depth of 1500 ft. (457 m) in only 5 mph (8 kph) wind! Also, one OASIS-7 windmill can pump as much as 150,000 U.S. gallons (568,000 liters) per day from a shallow well or river. That's why they are described as "incredible."

World Power Technologies, Duluth, MN
(Characterization: Wind turbine manufacturer, small systems.)

World Power Technologies offers a wide range of relatively low-power wind turbines ranging from 500W to 8500W in output voltages ranging from 12VDC to 240VDC. Available watt-age ratings are: 500W, 600W, 1000W, 1500W, 3000W, 4500W, and 8500W. Useful outputs are obtained in winds as low as 12 mph, peaking at about 25 to 30 mph. *Figure 4-13* and *Figure 4-14* show a fully assembled Whisper 1000 and the final assembly line at the World Power Technologies' Duluth, MN, plant.

The Whisper Mariner H500 was designed with sailboat owners in mind. It has extensive corrosion protection and polished stainless-steel surfaces. It is perfect for rooftop mounting

Figure 4-13. *Office manager Vicki Klein stands proudly next to World Power Technologies' Whisper 1000. This three-blade beauty can produce over 1000W of electrical power. Photo courtesy of World Power Technologies, Duluth, MN.*

Figure 4-14. *Production manager Ross Christenson performs inspections at the final assembly line. A Whisper 1000 generator is shown in the shipping container (foreground) and a variety of 1000s, 600s and 3000s can be seen on the line. Photo courtesy of World Power Technologies, Duluth, MN.*

to complement a photovoltaic system. The rotor consists of three epoxy-coated wooden blades available in 4 ft. and 5 ft. diameters. The Whisper 600 and 1000 watt wind turbines come with a two-blade epoxy-coated rotor but an optional three-blade rotor is available. The new Whisper 1000, shown in *Figure 4-13*, actually peeks at 1200W in a 30 mph wind. The new Whisper H1500 actually peeks at about 1800W in a 35 mph wind and has a 9 ft. diameter, three-blade rotor. The Whisper 3000 comes with a carbon-fiberglass two-blade, 14.8 ft. diameter rotor. A three-blade carbon-fiberglass rotor is also available. Also new for 1996 to the World Power Whisper line is the Whisper H4500. The H4500 comes with a 14.8 ft. diameter, three-blade rotor and reaches its rated power in 28 mph wind. The Whisper 8500 is designed to produce 8500 W in a 25 mph wind. It comes with a two-blade, 20 ft. diameter, carbon-fiber rotor and a three-blade rotor option is available.

All of World Power Technologies' wind turbines are built extensively with stainless steel parts and hardware. The overall design of their wind turbines is wonderfully simple and reliable. There are no gears, pulleys, or belts. The generator is driven directly from the turbine. This eliminates wasted power. In addition, the entire turbine-generator-tail assembly is designed to tilt back in high winds automatically relieving wind loading and preventing overspeeding. Extensive design development, both mechanical and electrical, has gone into these systems to make them reliable workhorses. Whisper wind turbines can also be used for water-pumping systems to provide water in remote areas of large ranches, etc. *Figure 4-15* shows a Whisper 1000 being used to supply a Texas rancher's cattle with water.

4-6. REVIEW QUESTIONS

1. What is the basic difference between a HAWT and a VAWT?
2. What is a synchronous inverter?
3. Is a Darrieus wind turbine a HAWT or a VAWT?
4. List two problems with chemical energy storage.
5. List two general problems and solutions to wind energy utilization.
6. What does AWEA stand for and who are they?

Check your answers from the *Answers to Questions* appendix at the back of this book.

4-7. WORLD-WIDE WEB SITES

The Author's Home Page

http://www.castlegate.net/personals/hazen/

Figure 4-15. *Here, the Whisper 1000 is used to provide energy to pump water for the cattle of this rancher in Wheeler, Texas. The direct pumping model PMP-2B supplies water for 120 head of cattle. Photo courtesy of World Power Technologies, Duluth, MN.*

Wind Energy Related

American Wind Energy Association
    http://www.econet.org/awea/
    http://solstice.crest.org/renewables/awea/index.html

Bergey Windpower Co., Inc.
    http://www.bergey.com

California Energy Commission
    http://www.energy.ca.gov/energy/html/directory.html

International Wind Energy Associations
    http://solstice.crest.org/renewables/wind-intl/index.html

Wind Energy in California
    http://energy.ca.gov/energy/wind/wind-html/wind.html

Wind Energy Web Sites
    http://www.strath.ac.uk/~cadx741/wind.html

World Power Technologies, Inc.
    http://www.webpage.com/wpt

5

HYDROPOWER:
HARNESSING WATER POWER

Oftentimes, as members of a very modern, high-technology society, we become smug in thinking alternative energy sources are modern concepts. However, as in the case of other alternative methods, using lakes, rivers, streams and the ocean as energy sources has been not just a concept, but an actual practice for hundreds, even thousands, of years. Ancient Greeks had water wheels to grind grain and crush grapes over 2,000 years ago. You will probably be as surprised to learn that the Chinese used crude, yet clever wave machines to crush iron ore as early as the 1200s. It is said that Leonardo da Vinci (1452-1519 AD), among his many other inventions, designed several wave machines. By the late 1800s and early 1900s, there was worldwide interest in extracting energy from the restless sea. Many turn-of-the-century wave machines are on file in patent offices around the world.

Then, from the early 1900s to about 1970, alternative energy hopefuls fell prey to the fierce competition offered by the coal and petroleum industries. The relatively high initial cost of large scale utilization of the oceans and energy sources forestalled any real progress. There just wasn't enough incentive to mother research and development. However, the oil embargo and subsequent oil shortage of the mid 1970s set a new stage upon which the quest for alternative energy sources would be reenacted. Many governments of the world set aside generous sums of money in emergency measures to pursue all viable energy sources. This quest for alternative sources was nurtured further in the wake of fear over the safety of nuclear power plants in the late 1970s and early-to-mid 1980s. Now there is the renewed fear of the reality that fossil fuels are running out. Somehow, some way, man knows he must tap clean inexhaustible alternative and renewable sources of energy. One of mankind's greatest and most imperative challenges is, and will be, to do just that. With 70% of the earth covered with oceans, it is little wonder that man has looked (and will continue to look) to the sea as a vast and inexhaustible source of energy.

5-1. HYDROPOWER: EXTRACTING ENERGY FROM WATER

Hydro: Greek, referring to water or liquid. The use of hydropower was as well known to ancient peoples, such as the Greeks, as it is to us today. Kinetic energy is contained in any fluid or liquid that is in motion. This motion, and kinetic energy, is the result of a difference in level or height of the source of the liquid and its point of energy extraction. This difference in height is referred to as *head*. The more head you have, the more kinetic energy there is. Another factor is the rate of flow in gallons per minute or liters per second. Again, the higher the flow rate, the more kinetic energy there is.

Hydropower is one of the least costly and most efficient means of doing work (conversion to mechanical and electrical energy). If you are fortunate enough to live near a continuously-running stream, river, or waterfall, you can produce large amounts of mechanical and/or electrical energy at minimal cost. Considering the amount of energy production and the initial low cost, hydropower is generally agreed to be the best choice. The catch is being fortunate enough to have a usable source of hydropower.

Extracting Hydropower

Extracting the kinetic energy from water that is in motion from a higher source can be as simple as installing a traditional water wheel. A water wheel can be installed at a narrow point in a stream, beneath a waterfall, or can be fed with an overhead trough or pipe. The wheel turns relatively slowly according to the natural flow of the water. However, the large diameter of the wheel produces a great amount of torque that can be geared up to a higher speed for turning a generator.

Where a water wheel is not practical, a water turbine can be used. In a situation where an elevated water source is nearby but there is no stream or point of natural energy extraction, piping can be installed to direct water downward to a turbine. Turbines are generally much smaller than water wheels and run at a much higher speed. This is because the water is forced through a jet nozzle at the turbine. The water velocity is used to push against the outer rim of the turbine wheel. Since the wheel spins at a high rate, gearing is not needed; the generator can be driven directly.

There are two major categories of turbines: *impulse turbines* and *reaction turbines*. Impulse turbines operate as high-speed water, via a jet nozzle, is sprayed at vanes or cups on the rim of a wheel. One of the most popular types of impulse turbine uses the Pelton wheel. The Pelton wheel is a wheel with cups all around its circumference. Another impulse type is the Turgo wheel, which has an outer rim tying all of the vanes together. Still another, but quite different, is the cross-flow turbine, which is more like a long roller than a wheel. The water flows through it, pushing against long, curved blades as it enters and exits.

Reaction turbines are those that have propeller-like rotors that are installed within a pipe. The water is forced through the blades causing them to turn. As with the impulse turbines, the speed of rotation is very high and can drive a generator directly.

Estimating Available Power

Estimating the amount of electrical power that can be generated from a given water source is fairly straightforward. Two things must be determined in order to make the power calculation: rate of flow, and head. The rate of flow can be determined in one of two ways:

1. Allow the flow from a pipe to fill a tank of known volume. Time how long it takes. Then calculate the rate of flow by dividing volume by time.

2. Drop a float in the middle of the stream and measure its velocity as it floats downstream. Measure the cross-sectional area of the stream. Approximate the flow from this formula: flow = 0.8 x velocity of float x cross-sectional area of the stream

Now you can calculate an estimate of the amount of available power in kW using this formula:

$$P(kW) = (F \times H)/102$$

where F is the rate of flow in liters per second and H is the head in meters. For example, let's say the flow is 20 liters per second and the head is 30 meters. $P(kW) = (20 \times 30)/102 = 5.88$ kW Remember, however, that this is the maximum available power. The actual electrical power you create will depend on the efficiency of your turbine system. That could be anything from 30% to 70%. Let's say you must generate 5 kW of electricity. If the efficiency of your system is, say, 40%, you will need available power of 5 kW/0.4 = 12.5 kW.

If you wish to use gallons per minute and feet to calculate available power, you can use this approximation:

$$P(W) \sim F \times H \times 0.18$$

where F is rate of flow in gallons per minute and H is head in feet.

DC or AC?

Most small and inexpensive hydroelectric systems produce DC. Many times, automotive-type alternators are adapted for this purpose. Usually a 1500 W impulse turbine of this type can be purchased for around $1500 U.S. However, storage batteries and an inverter are needed to complete the system. Hydroelectric systems that produce 120/240 VAC directly have a much higher initial cost because of the complexity of the turbine. The design must include means for maintaining constant frequency and voltage.

5-2. OCEAN WAVE POWER GENERATION

Anyone who has experienced the irresistible power of a crashing wave on a shoreline and the retreating undercurrent is very much aware of the ocean's great untamed energy. It is no marvel that man has been tempted over and over again to channel this energy for his own use. The principle behind wave power is actually very simple. The ocean offers up large masses of water that must return from whence it came. Mankind intervenes with a clever device that causes the water to work on its return. Thus, a wave is captured and the mass of the returning water, acted upon by gravity, is channeled and forced to turn a turbine shafted to a low-speed generator or alternator.

Wave energy systems fall in three basic classes:

1. Ocean based, or moored.
2. Near shore.
3. Land based.

Many moored wave energy systems are analogous to a huge floating funnel that has been moored in rough seas. The swelling waves are captured in the large mouth and funneled through hydroturbines down the venturi shaped shaft to the receding depths and undercurrents. Near shore designs involve large mechanical contrivances moored close to shore, to take advantage of incoming surface waves and outgoing undercurrents. The land based systems trap waves against the land mass and force the water to return through venturi tunnels containing hydroturbines. Older designs employed water wheels.

While this source of energy is free, it is far from being ideal. Waves are not always available in sufficient size, neither do they supply a constant flow of energy. They are cyclic and sporadic. Even though wave energy may be very adequate for some applications, such as milling, grinding, or crushing, it is not so compatible with the generation of electricity. The generation of electricity requires a smooth and continuous flow of mechanical energy. Therefore, the use of waves to generate electricity usually requires an intermediate form of energy for storage, such as storing wave pulsed DC in batteries and slowly converting it to AC using inverters (DC-to-AC converters).

5-3. OCEAN CURRENT POWER GENERATION

Although still mainly in the form of speculation and laboratory models, mankind has also envisioned using the strong, steady ocean currents as a dependable and vast source of energy for the generation of electricity. The basic concept entails the use of large numbers of giant underwater rotary sails similar to windmills. The huge turbines would be anchored firmly, turning slowly with irresistible power. The electricity would be cabled to shore.

Like many ocean energy systems, this one has many significant disadvantages. The growth of barnacles and other creatures on the apparatus will unbalance the blades and impair operation. Many people are concerned for the safety of larger sea creatures that may venture into harm's way. Then, there is the very practical problem of cabling the electrical energy to shore from hundreds of miles at sea; an expensive problem, though not impossible. One possible solution is the conversion of ocean energy to chemical energy by the production of hydrogen through electrolysis. The ocean generators would be used to separate water into oxygen and hydrogen. The hydrogen would then be compressed and contained for transport to land for use as fuel in large generating plants.

5-4. OCEAN TIDE POWER GENERATION

There are some locations in the world where the ocean level rises 30 to 50 ft. (9 to 15 m.) during high tide. A these locations include the Bay of Fundy in Canada adjacent to Maine; the Severn Estuary in Great Britain; the Rance Estuary in France; and the Gulf of California in Mexico. These locations present tremendous opportunities to trap the water in landlocked bays and use it to drive hydroturbines as it returns to the ocean at low tide. This source of energy is an attractive one since the tides are very predictable and reliable. However, it is a source of energy that can only be tapped at certain geologically-suitable locations around the world — locations where large masses of tidal water can be trapped and slowly returned to the ocean. For this reason, it is not available to everyone in the same way that wind, sun, waves, and currents are available.

As you might expect, the use of tidal power is not a new idea but dates back to the grinding mills of the Middle Ages in Europe and the colonial period of the United States. In more modern times, the French completed a 240,000 kW tidal system in the Rance Estuary in 1967. The Rance installation produces over 500,000,000 kWh of electricity per year. The Soviets completed a 400,000W experimental system at Kislayaguba, north of Murmansk, in 1969.

Other prime areas have been studied and may be under development at this time. A two-basin tidal-energy installation is now being considered for the Bristol Channel of Great Britain. It has been estimated that the site would have the ability to produce 12 million, million watts of electricity. That's 12,000,000,000,000 watts! It would take care of 5% to 7% of Britain's electricity needs. British authorities believe recent improvements in turbine design and construction techniques make such projects comparable in cost to nuclear power installations.

Some concerns that exist in regard to tidal plants include: What impact will the installation have on the local environment during construction? What impact will the installation have on the local environment *after* construction is complete? Also, the initial installation costs are very high. Nevertheless, it has been shown that this alternative source of energy, in the long run, is cheap, reliable, predictable, and clean. Unfortunately, there are limited numbers of geographical locations suitable for such an installation.

5-5. REVIEW QUESTIONS

1. Briefly explain how ocean waves can be used as an energy source to generate electricity.
2. Explain how energy collected from ocean currents might be transported to land for use in electrical generating plants.

Check your answers in the *Answers to Questions* appendix at the back of this book.

5-6. WORLD-WIDE WEB SITES

The Author's Home Page

http://www.castlegate.net/personals/hazen/

Regarding Hydropower

Alternative Energy Engineering, Inc.
   http://www.alt-energy.com/aee

Alternative Energy Sources Course, U. of Oregon
   http://zebu.uoregon.edu/1996/phys162.html

Centre for Alternative Technology
   http://www.foe.co.uk/CAT/index.html

Mr. Solar's Home Page
   http://www.netins.net/showcase/solarcatalog/

Solstice
   http://solstice.crest.org/

6

GEOTHERMAL ENERGY

The earth is a great, multilayered ball, most of which is extremely hot. The outer layer, called the *crust*, is relatively thin, solid, and cool. Its thickness varies from 12 to 37 mi. (20 to 60 km.) on land and as little as 3 mi. (5 km) under the oceans. Below the crust is a very thick, hot layer called the *mantle*. Most of the mantle is solid, yet very hot, and reaches 1800 mi. (2900 km) from the crust to the next inner layer, called the *outer core*. The outer core is extremely hot molten rock called *magma* (5400° to 9000° F or 3,000° to 5,000° C). It is the magma that finds its way through the mantle under great pressure to the surface as a volcano. The inner core occupies the center of the earth and has been found to be solid under unfathomable pressure and temperature. The entire core, including the liquid outer core, is about 2200 mi. (3500 km) in radius, making it larger than the planet Mars. The heat contained by the earth in its core and mantle represents a nearly limitless source of energy. (Limitless in a practical sense, since the cooling process is measured in millions of years.) It is no wonder that mankind is interested in tapping this vast storehouse of energy. Here we will explore some of the methods and means by which this energy can be (and is) extracted from the earth.

6-1. HYDROTHERMAL GEO-ENERGY

Hot Water Hydrothermal Energy

One method used to extract thermal energy from the earth is referred to as *hydrothermal*. As shown in *Figure 6-1*, hydrothermal energy is manifested in two general ways: (1) as hot springs and geysers, and (2) as dry steam. From deep in the earth, thousands of meters into the crust, water reservoirs are superheated by the hot mantle or, in some cases, seeping magma. The highly pressurized hot water seeks passage through fissures in the crust to finally escape at the surface. Hot water hydrothermal sources are generally used in direct heating applications such as homes, fish ponds, and various buildings. A grand example of this is found in Reykjavik, the capital of Iceland, where hot water hydrothermal energy is used to directly heat the homes of approximately 100,000 people.

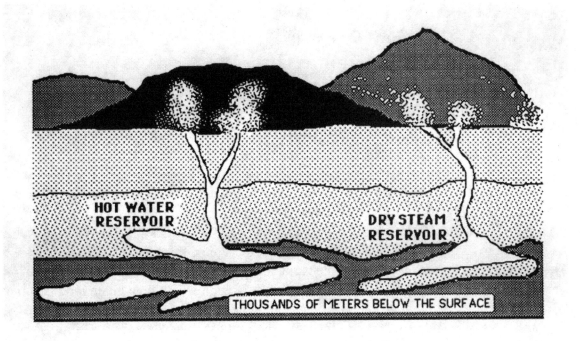

Figure 6-1. *Hydrothermal energy source*

Dry Steam Hydrothermal Energy

Dry steam is produced when water trapped deep in the earth is superheated by the hot mantle or magma. High-pressure steam forces its way upward and is released at the surface as an extremely hot gas having little or no liquid content. Dry steam sources are generally used to drive steam turbine generators for the production of electricity. The Lardarello Field in Italy is one example of dry steam electricity production established in 1904 and still running strong today with a total capacity in the neighborhood of 600 MW. The largest dry-steam hydro-thermal electricity generating plants in the world are owned by Pacific Gas & Electric (PG&E) and others in California. These plants are located about 72 miles northeast of San Francisco, CA, and have a total capacity in excess of 1400 MW. California is the leading state in the utilization of geothermal energy for the production of electricity. California, Nevada, and Oregon have the largest number of geothermal hot beds in the United States.

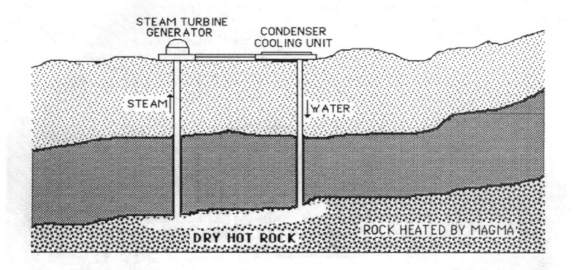

Figure 6-2. *Hot rock is used to produce steam.*

6-2. HOT-ROCK GEO-ENERGY

Hot-rock geo-energy is closely related to hydrothermal energy in that extremely hot rock deep in the earth's crust is used to superheat water, turning it into highly pressurized steam. In this case, however, the rock is naturally dry and water is pumped down from the surface to be super heated. *Figure 6-2* illustrates the process. Water is pumped deep into the crust to a dry, hot rock bed (400° to 1800° F or 200° to 1000° C). The resulting steam is channeled to the surface to drive a steam turbine generating system. The steam is condensed in a massive cooling unit and recycled to the deep hot rock bed. Ideal locations for hot rock installations are those that have large hot rock beds at depths no greater than 3 or 4 km, through a crust that permits easy drilling. In most cases, suitable hot rock beds are found in volcanic regions.

6-3. GEOPRESSURIZED WATER ENERGY

Geopressurized water energy is another geothermal source of energy that is being pursued by geologists the world over. As shown in *Figure 6-3*, geopressurized water is water that is trapped in huge chambers or reservoirs deep in the earth's crust. The water is held under great pressure and temperature. Once the reservoir is located, it must be successfully tapped,

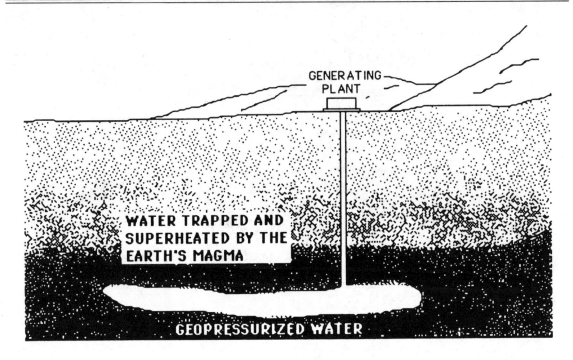

Figure 6-3. *Geopressurized water.*

a task easier said than done. If the reservoir is tappable, hot water, steam, and many natural gases are usually be available for heating and generating electricity. Naturally, it must be predetermined that the reservoir is sufficient in size to justify the costs of labor, equipment, and facilities.

6-4. ADDITIONAL GEOTHERMAL SITES

Naval Weapons Center, China Lake, CA
(Geothermal Program Office Naval Air Weapons Station
China Lake, CA 93555-6001 (619) 939-2700)

Geothermal steam turbine facilities at the Naval Air Weapons Station in China Lake, CA, produce over 250 MW of electricity, saving the Navy and U.S. citizens millions of dollars every year. It has been estimated that this installation can meet the energy needs of a million people with a savings of 3,825,000 barrels of oil or 875,000 tons of coal every year!

The Geysers in California
(Information supplied courtesy of Pacific Gas & Electric Company)

The Geysers, located about 30 miles northeast of Healdsburg, CA, and about 72 miles north of San Francisco, covers about 40 square miles and is considered the largest geothermal dry-steam hydrothermal location in the world. The mountainous area was named the Geysers only because of large plumes of steam escaping from the ground in various locations. The plumes of steam somewhat resemble actual hot-water geysers. In reality, it is a misnomer since a steam plume is technically not a geyser — it is a *fumarole*. With that fact acknowledged by all, it remains the "Geysers."

The Geysers was discovered in 1847 by an explorer/surveyor by the name of William Bell Elliot. While hiking through the mountains on a grizzly bear hunt, Elliot happened upon what was to him a startling sight. What at first appeared to be smoke billowing out of the ground with a fire and brimstone odor, he soon discovered was very hot steam. Upon report of his discovery to his friends, Elliot said, "I thought I had come upon the gates of hell." By 1863, the area had become somewhat of a tourist attraction with such notable visitors as Ulysses S. Grant and Theodore Roosevelt.

Early attempts to harness the power of "hell's gates" started as early as 1922. Unfortunately, turbine and materials technology was not yet far enough advanced to build a reliable, long-lasting plant. With the abundance of hydroelectricity to meet energy needs, the idea of harnessing geothermal was put off until the mid 1950s. On September 25, 1960, Pacific Gas & Electric (PG&E) went online with an 11 MW geothermal plant, marking the beginning of the practical utilization of geothermal energy for the entire nation.

Since that day in September, 1960, over 600 wells have been drilled to depths averaging 8500 ft., with some two miles deep. The 355°F steam emerging from these wells provides turbine pressures ranging from 60 to 100 psi. Enough steam is available to generate millions of watts of electricity. Before the steam is piped into the special turbines, it must be cleaned in a centrifugal separator to remove rocks and debris that could damage the turbines.

Once energy has been extracted from the steam by the turbines, it is condensed and pumped to cooling towers. From there, the water is gravity-fed into the earth through deep injection wells. In this way, the water is recycled and returns again as steam. This is vital toward continuing the cycle. Scientists and engineers are learning this and other management techniques in order to greatly extend the useful life of the site. The entire Geysers reached a peak of 2000 MW in 1987. Today, the total output has dropped to a little over 1400 MW. All total, geothermal energy production at the Geysers saves the burning of some 13 million barrels of oil per year.

Other entities involved in steam production and electricity generation, in addition to PG&E, are Magma Power Company, Thermal Power Company, Union Oil Company (UNOCAL), Calpine Corporation, Northern California Power Agency (NCPA), Central California Power Agency (CCPA), Sacramento Utility District (SMUD), and Santa Fe International.

Geothermal at the 1996 Summer Olympics in Atlanta, GA

Geothermal heat pumps, used for space heating/cooling and hot-water heating, are in operation at the Education and Training Center and in a newly-renovated building at Georgia Tech in downtown Atlanta, GA. This was sponsored by the DOE as a demonstration project for the 1996 Summer Olympics. Using the ground as a heat source or sink, these systems demonstrate that geothermal heat pumps are more efficient and have lower life-cycle costs than conventional air-source heat pumps (such as heat-pump air conditioners). *Figure 6-4* shows workers drilling holes for heat exchanger pipes for the system as it was being installed before the summer games.

6-5. ENVIRONMENTAL CONCERNS

Air Pollution

While it might first appear as though geothermal energy is very clean and will have little or no impact on the surface environment, there are some very real environmental concerns. First, geothermal energy sources can contribute to air pollution. Water and steam are not the only products extracted from the earth's crust. Many hazardous gases such as hydrogen sulfide, carbon dioxide, methane, and ammonia can be released in the process. Methods have been and are being developed to capture these secondary products and put them to useful applications.

Water Pollution

A second concern is in the area of water pollution. Subterranean water contains many minerals, including large quantities of salt. If this water is permitted to mix with surface waters, plants and animals will be destroyed and the local ecological system will be disrupted. The

Figure 6-4. *Here, workers are drilling holes for heat exchanger pipes for a geothermal heat-pump system in a newly-renovated building on the Georgia Tech Campus. Photo courtesy of the U.S. Department of Energy, National Renewable Energy Laboratory, and Craig Miller Productions.*

solution to this problem seems to be purification or reinjection of subterranean waters back into the underground reservoirs.

Surface Ground Effects

A third concern is that of ground subsidence. As water is extracted from the crust, and subterranean pressure is relieved by exhausting steam, there may be (and in some cases has been) a tendency for the ground to crack or sink (subside). Careful geological studies must be performed to insure a solid crust structure before energy extraction begins.

U.S. Department of Energy (DOE)

In the United States, the Department of Energy (DOE) has established a thorough program of checks and balances to insure environmental safety at every stage of geothermal development. With careful planning and utilization of the latest technological advancements, these environmental concerns can be (and have been) addressed with great success. For more information on the U.S. Department of Energy, see the last chapter of this book.

6-6. REVIEW QUESTIONS

1. What are the two forms of hydrothermal geo-energy?
2. How is hot-rock geo-energy different from hydrothermal?
3. List two geothermal environmental concerns and a solution to each.

Check your answers in the *Answers to Questions* appendix at the back of this book.

6-7. WORLD-WIDE WEB SITES

The Author's Home Page

http://www.castlegate.net/personals/hazen/

Relating to Geothermal Energy

CREST's Intro to Geothermal
    http://solstice.crest.org/renewables/geothermal/grc/index.html

Alan Glennon's Geyser Page
    http://www.wku.edu:80/~glennja/pages/geyser.html

BRIDGE - British Mid Ocean Ridge Initiative
    http://www.nwo.ac.uk/iosdl/Rennell/Bridge/

Coso Geothermal Project
250 MW Geothermal power plant in California
    http://www1.chinalake.navy.mil/Geothermal.html

Fourth International Meeting
Heat Flow and the Structure of the GeoSphere, June 10-16, 1996
    http://www.eps.mcgill.ca/~hugo/heat.html

Geothermal Education Office
    http://www.ensemble.com/geo

Geothermal Energy in Iceland
    http://www.os.is/os-eng/geo-div.html

Geothermal Exploration in Korea
    http://www.kigam.re.kr/env-geology.html

Geothermal Heat Pump Consortium
    http://www.ghpc.org/index.html

Geothermal Heat Pump Initiative in the U.S.
    http://www.eren.doe.gov/ee-cgi-bin/cc_heatpump.pl

Geothermal Resources Council (USA) Library & Information
    http://www.demon.co.uk:80/geosci/grclib.html

Geothermal Section
Course in Renewable Energy (Physics 162)
Part of the Electric Universe Project at Univ. of Oregon
    http://zebu.uoregon.edu/ph162/l18.html

Geothermics
International Journal of Geothermal Research and Its Applications
    http://www.elsevier.nl/catalogue/SA2/230/03700/03770/389/389.Html

International Geothermal Association
    http://www.demon.co.uk:80/geosci/igahome.html

Rotorua, New Zealand, Geothermal Areas
    http://www.akiko.lm.com/NZ/NZTour/Rotorua/Geothermal.html

Stanford University Geothermal Program
    http://ekofisk.stanford.edu/geotherm.html

U.S. Department of Energy
Energy Efficiency and Renewable Energy Network
    http://www.eren.doe.gov/ee_renen-geo.html

7

NUCLEAR ENERGY

We are getting ready now to take a look at a form of alternative energy that once held great hopes for replacing our dependence on fossil fuels to generate electricity. That, of course, is nuclear energy. When the first reactors were being built 50 years ago, many believed nuclear energy would be the ideal, nearly utopic means of generating electricity. It was thought that it would be very inexpensive and nonpolluting. As if in a gold rush, utilities all over the world made plans to build nuclear reactors. In the early 1970s, the U.S. Atomic Energy Commission (replaced by the Nuclear Regulatory Commission (NRC) in 1974) predicted that there would be over 1,000 nuclear power plants in the United States by the year 2,000. By the late 1970s, the realities of nuclear-energy production began to set in. Instead of proceeding with plans to build more nuclear plants, utilities began canceling such plans with, no new requests for nuclear plants in the U.S. since 1978. The industry was plagued with problems that had not been predicted, including cost overruns in construction, early deterioration, high maintenance costs of containment structures and fixtures, and a national nuclear waste repository that was first planned in 1957 but to this day has not materialized. Existing plants, numbering 110 in the U.S. and over 400 worldwide, are now scrambling for answers to what has now become a vexing problem. Customers are being charged extra for energy usage, and the money is being set aside for a national repository someday. Meanwhile, some of the customer surcharge is being used to pay for waste storage right on site since there is nowhere else to take it! For additional facts and figures, see the August 1995 issue of *Popular Science Magazine* and investigate the topic on the Internet.

"The technology unleashed by bomb builders was supposed to bring us cheap, clean energy. But operating nuclear power plants has become so expensive that many are quietly mutating into radioactive waste dumps."

— Dawn Stover, "The Nuclear Legacy," *Popular Science Magazine*, August 1995, pg. 52.

The Washington International Energy Group, an energy consulting firm, released the results of a survey of power company executives and independent power producers in February of 1996. The survey showed that only 8% of those who responded to the survey said they believed there would be a resurgence in the utilization of nuclear energy while an overwhelming 80% said there would not. Also, only 2% said they would consider ordering a new nuclear power plant while 89% said they would not. These results represent a significant drop from surveys of previous years and strongly indicate the attitude of the utilities. More and more utilities are looking toward renewable alternative energy sources. A copy of the survey may be obtained from the Washington International Energy Group by calling (202) 331-9820.

As you have seen above, many today no longer view nuclear energy as an "alternative." We now know that the costs and risks are very high for the traditional fission reactors. That is why there is a renewed and continuing determination to utilize safer alternatives most frequently referred to as "renewables." Renewable energy sources such as solar, wind, and hydro carry with them no inordinate threats to the environment beyond perhaps changing the appearance of the landscape. Still, even if no new nuclear fission plants are built, it will be some time before we can rid ourselves of the current nuclear plants and waste what we have already created. Some scientists say that it will take 10,000 years for the spent fuel to deteriorate to a safe level while others say only 500 years. In spite of this, we find ourselves strongly dependent on nuclear energy. Over 20% of the electricity generated in the U.S. is nuclear in source. Not only is that hard to replace, but our energy needs continue to increase and world oil supplies continue to diminish.

In spite of the negative aspects of nuclear energy, it is still in use as an alternative to fossil fuels, and nuclear research continues. Also, countries such as China and India, which are emerging technologically and economically, will definitely be building new plants to assist with their rapidly expanding energy needs and to take the place of declining oil supplies. Many are looking more toward *breeder reactors* for the more immediate future, and *fusion* for the more distant future. We in the United States may find ourselves once again building nuclear fission plants to take up the additional demand and to compensate for diminishing oil and coal. That would be a tough decision for a nation desiring to move away from nuclear fission.

At present, the number of nuclear plants in the U.S. continues to decline due to decommissioning. In light of the present and *possible* future use of nuclear energy, an investigation into the science of nuclear energy is in order. In this chapter, we will explore the world of nuclear energy to see how nuclear reactions occur; how nuclear energy is converted to electricity; the differences between various nuclear reactors; and to peer wistfully into the future with a brief discussion of nuclear fusion.

7-1. IT BEGAN WITH ENRICO FERMI

Enrico Fermi (1901 - 1954)

Enrico Fermi was an Italian-born physicist who immigrated to the United States in 1938. As a physicist, he did much research on nuclear processes and uranium. Fermi won the Nobel Prize in Physics in 1938 for his research. He was the first man to split the atom in a 1934 experiment. At that time, he thought he had created a new element when, in fact, he had split the uranium atom creating two known elements. Fermi became a professor of Physics at the University of Chicago in 1942 and there led a team of physicists in performing the world's first controlled nuclear fission chain reaction.

On December 2, 1942, mankind officially entered the atomic age as the first controlled nuclear reaction was created by a team of dedicated scientists led by Enrico Fermi. Their experiment was staged in the humble environs of a racquetball court underneath the west spectator stands of Stagg Field at the University of Chicago. Dr. Fermi and his team of scientists proved that a nuclear chain reaction was possible and that it could be controlled. At the time, their findings were held as top secret by the U.S. Government since this new technology was needed for the development of the atom bomb used to end WW II. However, it was the intent and hope of all of the participating scientists that nuclear energy be put to use in service to man in peacetime surroundings. In 1955, the first nuclear powered electricity generating station was placed in service in Arco, Idaho.

Since 1955, hundreds of nuclear power plants have been constructed around the world in an effort to supply the ever-increasing demand for electricity and serve as a substitute for coal, fuel oil, and natural gas. Without any doubt, nuclear energy stands head and shoulders above all other forms of alternative energy in answering the increasing energy demands. However, there is also no doubt that this source of alternative energy is the most feared. The threat of accident or disaster is real, though the probability is low. Society can't help but be leery in light of accidents like Three-Mile Island in Pennsylvania and, most tragically, Chernobyl in Russia.

BOILING WATER REACTOR (BWR)

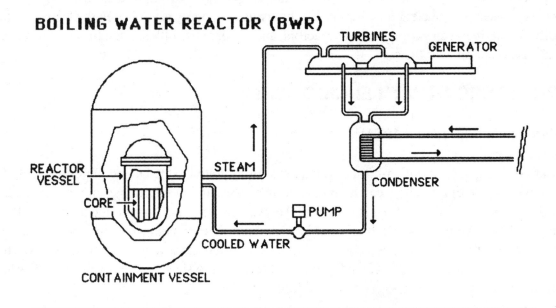

PRESSURIZED WATER REACTOR (PWR)

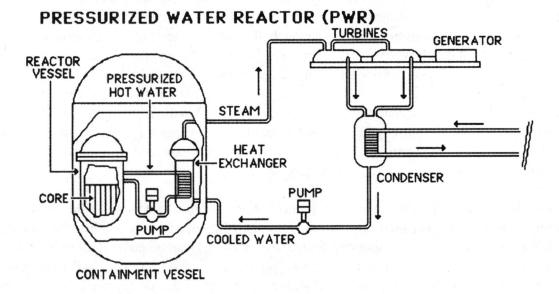

Figure 7-1. Light water reactors (LWRs).

7-2. LIGHT WATER NUCLEAR REACTORS (LWRs)

BWRs and PWRs

Nearly all of the nuclear reactors in existence today are those that fall under the broad heading of light water reactors (LWRs). LWRs use pure water as the main coolant for the reactor itself. *Figure 7-1* shows the overall system diagrams for the two main types of LWRs: the boiling water reactors (BWRs) and the pressurized water reactors (PWRs). As you can see, LWRs are actually steam turbine generating plants. The nuclear reaction either directly or indirectly superheats water, converting it to steam to be used in powering the turbines. In boiling water reactors, the water is converted directly to steam, while in the pressurized water reactors, a pressurized water system carries large quantities of heat from the reactor to a heat exchanger where water in a secondary system is converted to steam.

U-235 Fission Theory

In the core of the nuclear reactor, a large and steady flow of heat is generated as a nuclear chain reaction occurs. As stated above, the tremendous quantity of heat is used to convert

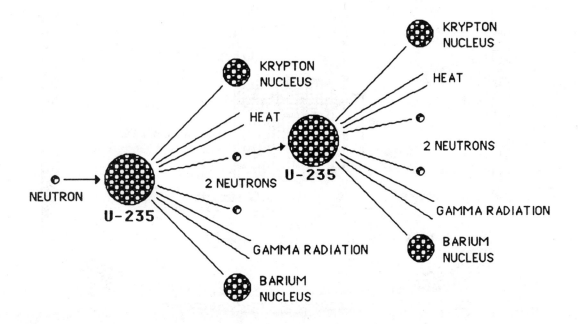

Figure 7-2. *U-235 nuclear chain reaction.*

water to steam in order to power a steam turbine that drives the huge electricity generator. How does this chain reaction occur, and what are the materials used in the reaction? In today's reactors, the nuclear fuel is uranium 235 (U-235), an isotope of uranium that has 235 neutrons in the nucleus of each atom. The nucleus of the U-235 atom is unstable enough to split when bombarded with an accelerated neutron from another atomic nucleus. The neutron is a very heavy subatomic particle. *Figure 7-2* illustrates this reaction. The splitting of the atom's nucleus, and therefore the atom, is known as *fission*. During the reaction, fission products are generated. These fission products include a newly-created krypton atom; a newly-created barium atom; two free and highly-accelerated neutrons; gamma radiation; and heat. In a chain reaction, the two liberated neutrons collide with other U-235 atoms which, in turn, fission and liberate more neutrons to continue the reaction. All the while, large amounts of heat are liberated and absorbed by the water coolant surrounding the fuel assemblies.

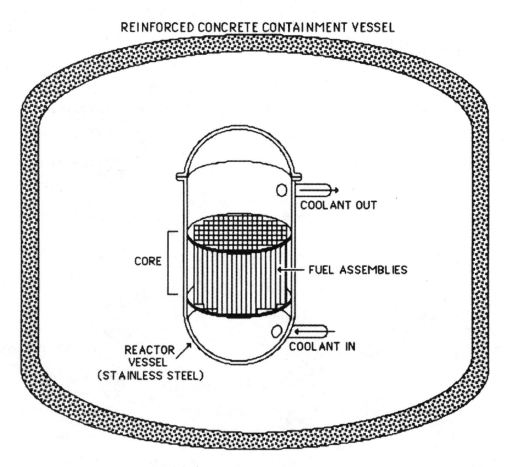

Figure 7-3. Reactor core assembly.

Figure 7-4. A fuel assembly is being loaded into a reactor core.
Photo courtesy of Nuclear Energy Institute, Washington, DC.

Reactor Core Assembly

As shown in *Figure 7-3*, the core of the reactor is contained in a heavy stainless steel chamber called the *reactor vessel*. *Figure 7-4* shows a reactor vessel with its lid removed for fuel loading. A fuel assembly is being moved into position. The reactor vessel is further enclosed in a larger reinforced concrete structure called the *containment vessel*. The purpose of the containment vessel is to trap radiation and radioactive substances and prevent them from being released into the environment in the event of a system failure or major accident.

Within the reactor vessel is the *reactor core*, composed of 200 to 300 fuel assemblies, as shown in *Figure 7-4*. Each fuel assembly is made up of a large bundle of individual long

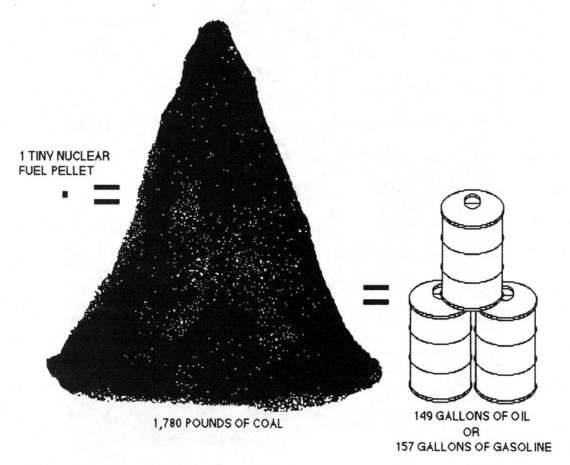

1 TINY NUCLEAR
FUEL PELLET

1,780 POUNDS OF COAL

149 GALLONS OF OIL
OR
157 GALLONS OF GASOLINE

Figure 7-5. *One nuclear fuel pellet contains the same amount of energy as 1,780 pounds of coal, 149 gallons of oil, or 157 gallons of gasoline.*

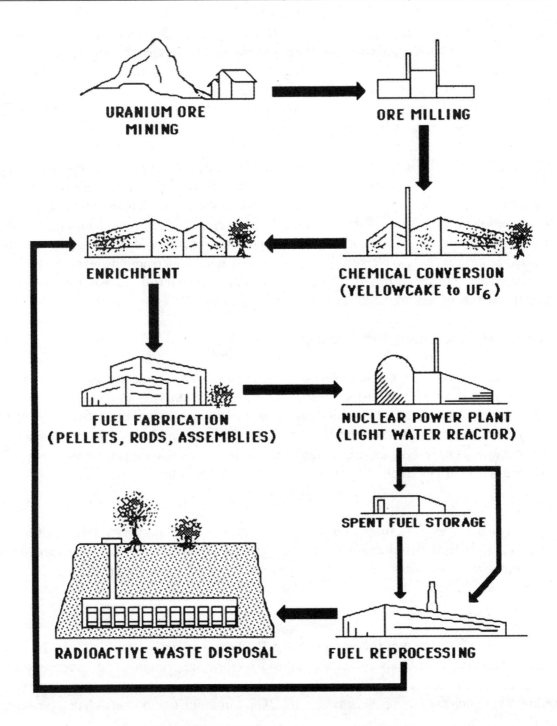

Figure 7-6. *U-235 production and cycling.*

tubular fuel pins, commonly called *fuel rods*. There are typically 236 fuel rods per fuel assembly. A fuel rod contains approximately 250 individual fuel pellets, each about 1 cm. in diameter and 1.5 cm. long. Each fuel rod is sealed, and liquid coolant (water) is permitted to flow between them in the fuel assemblies. It is interesting to note that one tiny fuel pellet contains the same amount of energy as 1,780 pounds of coal, 149 gallons of oil, or 157 gallons of gasoline. This is illustrated in *Figure 7-5*.

Interspersed among the fuel rods in the fuel assemblies are *control rods*. The control rods are made of materials that are able to trap and absorb neutrons. By maneuvering the control rods in or out of the core, the number of effective neutrons can be controlled; therefore, the rate of chain reaction is controlled. If the chain reaction is being sustained at a constant rate, the reaction is said to be *critical*. If the power level and chain reaction is decreasing, the reaction is said to be *subcritical*. Conversely, if the power level and chain reaction are increasing, the reaction is said to be *supercritical*. The control rods must be carefully adjusted in the core to maintain the rate of reaction at the critical point.

U-235 Fuel Production and Cycling

The uranium isotope U-235 is not available in great abundance in nature. Most of what is extracted from the earth is non-fissionable U-238. In fact, only about 0.7% of the mined uranium is the fissionable U-235. This concentration is far too low to be usable in a nuclear reactor. However, through a very involved process of mining, milling, chemical conversion, enrichment, and fabrication, the crude uranium ore is made suitable for reactor use. *Figure 7-6* illustrates the various steps in the fuel production and recycling process.

Mining
The first step in the fuel production process is the mining of the very crude and impure uranium ore. In the richest uranium fields, there is only about 0.2% uranium contained in the mined ore. This low concentration of uranium must be greatly improved in the long production process.

Milling
From the mine, the uranium ore is sent to a mill where it is further purified. At the mill, unwanted inorganic materials are removed through crushing, grinding, and various separation processes. The end result of this milling process is a substance referred to as "yellowcake." Yellowcake contains a large concentration of U_3O_8, a uranium compound rich in a variety of uranium isotopes.

Chemical Conversion

Unfortunately, yellowcake contains very little of the fissionable U-235. In fact, about 99.3% of the compound is U-238, a non-fissionable isotope, leaving 0.7% as U-235. The concentration of U-235 must be increased in order to be useful as a reactor fuel. The next step in the processing is to convert the U3O8 to a uranium compound that can more readily be enriched. This step is known as the *conversion stage*, in which U3O8 is converted to UF_6 (uranium hexaflouride).

Enrichment

The UF_6 is then sent to an enrichment plant where the concentration of U-235 is increased. During this stage of processing, two distinct compounds with very different concentrations of U-235 are produced. The compound with the higher concentration will contain about 3% U-235 and will be used for reactor fuel. The lower concentration compound will contain only about 0.2% U-235 and will be stockpiled for other purposes. This lower concentration material is referred to as *depleted uranium* or *tails*.

Fabrication

Finally, the enriched compound is sent to a fabrication plant. At the fabrication plant, the uranium fuel (enriched UF_6) is converted to UO_2 (uranium dioxide), a very chemically stable substance. The UO_2 is then compressed and baked into the form of the small fuel pellets. The pellets will then be loaded into *zircaloy tubes* (a zirconium alloy), forming the fuel rods.

Recycling

One fuel load will last anywhere from 18 months to 3 or more years in the core of a reactor. When the fuel reaches a level of inefficiency it is said to be *spent*, and must be removed and replaced. The spent fuel is then stored or (theoretically) recycled. The recycling process involves spent fuel being sent first to a reprocessing plant, where radioactive fission products are removed and residual U-235 and Pu-239 (plutonium) are recovered. The recovered fissionable materials are then sent back to the enrichment plant for further processing and enrichment. The problem here is that in 1977, reprocessing was outlawed because the plutonium that resulted was concentrated enough to be used in nuclear weapons. Thus, spent fuel can only be stored, not reprocessed. It was hoped that the U.S. could lead the world in banning the production of plutonium as a deterrent to nuclear proliferation. It didn't work. No other nuclear-capable country took our lead. Politicians now view the topic as a "hot potato," and it is doubtful that reprocessing will be reinstated in the U.S.

Waste Disposal: A Serious Problem

So, what happens to all of the spent fuel from all those hundreds of nuclear reactors around the world? It has turned into a real problem. A typical nuclear reactor in the U.S. produces 20 metric tons of spent fuel every year, and estimates are that by the year 2003, the U.S. will have 48,000 metric tons of spent fuel with nowhere to put it. The national repository for spent fuel discussed decades ago may not be ready until beyond the year 2010. In the meantime, nuclear plants are stockpiling their own spent fuel and (in some cases) ceasing operations early for lack of safe storage. On top of all this, the U.S. has agreed to take the spent fuel of some other countries once space has been made available—a dubious promise. Proposals are even being discussed with Native Americans to place spent fuels on their reservations since these are "sovereign nations." The next problem is that of legally and safely transporting spent fuel to the reservations. Legislators, pressured by anxious constituents, have threatened to vote against such transport while many Native Americans are pressing in favor of the idea based on financial gain. In all fairness, the DOE and many scientists believe the spent fuel can be transported and stored at temporary locations safely with the use of carefully designed containers and waste solidification techniques that can withstand severe mishap.

So, what do we do with the waste for the long run? Shoot it to the moon? Abandon it in space? Bury it in the oceans? Hide it under a mountain? Yes, that's it, a mountain. In 1987, Congress decided that Yucca Mountain, 100 miles northwest of Las Vegas, would be a possible best choice. Since then, scientists have been drilling, boring, and testing in every conceivable way on, in, and under the mountain to determine its suitability. The cost of all this probing to the American taxpayers is around $1 million PER DAY! The testing is expected to go on for a decade or more. Care to do the calculations? In the meantime, the nuclear waste stays at home with nowhere to go. That's a big reason why more and more people and utilities are deciding that fission-reactor nuclear energy is no longer a desired or practical form of alternative energy!

7-3. BREEDER REACTORS (BRs)

Comparing Breeder Reactors and LWRs

As shown in *Figure 7-7*, breeder reactors are different from LWRs in several very obvious ways. First, liquid sodium is used as a reactor coolant and heat exchange fluid instead of water. Sodium is a metal, and is far more efficient at heat absorption and heat transfer than

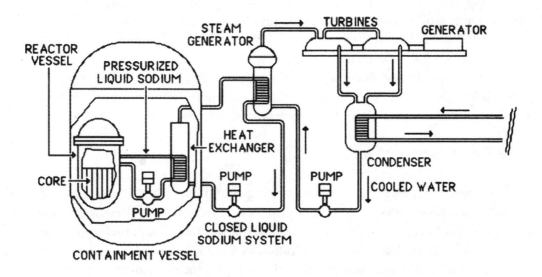

Figure 7-7. *The liquid metal fast breeder reactor.*

water. The increase in efficiency allows the reactor to be operated at a much higher temperature and lower coolant pressure. Because of the coolant used, these reactors are often referred to as *liquid metal fast breeder reactors* (LMFBRs). A second difference between the LMFBR and the LWR is the type of fission fuel used. Breeder reactors use U-238 and plutonium (Pu-239) instead of U-235. Although plutonium is not found in nature, it is a natural byproduct of LWRs and will reproduce itself (breed) in a nuclear chain reaction. That brings us to the third important difference between LMFBRs and LWRs: breeder reactors can actually breed more plutonium than they consume, a fact that makes breeder reactors very attractive. The catch here is that these reactors are much more costly to build than LWRs. Also, there is the political issue of producing plutonium. At present, there are only five breeder reactors in the world: the French have built two, the U.S. has a small one, the British have one, and the Russians have one. The Japanese are presently building one.

Pu-239 Fission Theory

The overall plutonium chain reaction is similar to the U-235 reaction in that U-238 and plutonium nuclei are split (fissioned) and fission products consisting of heat, liberated neutrons (more than 2), radiation, and two new elements are created. However, in the plutonium reaction, accelerated neutrons will accomplish two tasks: (1) nuclei will be split, and (2) more plutonium is created. How is it possible for something that is being destroyed to actually thrive and breed? Among the fissioning plutonium nuclei are many U-238 nuclei. Some of the liberated and accelerated neutrons will collide with the U-238 nuclei and will actually form plutonium Pu-239 (238 neutrons + 1), instead of the U-238 fissioning. Thus, during the chain reaction, more fissionable material is being produced from the absorption of liberated neutrons in U-238 nuclei. The plutonium chain reaction is very rapid due to the breeding of fuel and the increased number of liberated neutrons.

Pu-239 Fuel Production and Cycling

Pu-239 fuel production does not start with mining. Plutonium is not found in nature. Rather, plutonium production starts with recycling spent fuel from LWRs. Recall that some plutonium is produced during chain reaction in light water reactors. Spent fuel from these reactors can be processed to enhance the concentration of Pu-239 among U-238. Also, spent fuel from breeder reactors is rich in Pu-239 and can be reprocessed over and over. Thus, an unlimited supply of Pu-239 can be perpetuated in the breeding process from the very common U-238 uranium isotope.

Breeder Reactors Today

Breeder reactors are more complicated and more expensive to construct. For these reasons, breeder systems are not in common use around the world. However, many countries have been pursuing the advantages of this alternative energy source. Nations such as France, Russia, West Germany, Britain, and the United States have been in the forefront of research and development with one or more demonstration plants in operation in each country. The breeder reactor seems to be the next practical step for mankind in pursuing the elusive ideal alternative energy source. In the very long term, breeder reactors will probably be built in large numbers since they can produce 140 times as much energy from a kilogram of uranium than an LWR. The greatest barriers now seem to be extremely high construction costs and political issues over the production of plutonium.

7-4. QUEST FOR FUSION

Fusion is another type of nuclear reaction in which the nuclei of two atoms are forced together to form an atom of a different element. The nuclei are said to be fused together forming one larger nucleus. In the fusion process, some of the nuclei matter is converted into heat energy, which turns out to be very important in sustaining the reaction. As you can see, this is quite the opposite of nuclear fission reactions in which nuclei are divided. There are actually many differences between fission and fusion. The following comparisons will reveal the strengths and weaknesses of nuclear fusion technology.

Comparing Fission and Fusion

1. Fission is well understood and has been utilized for many decades to produce electricity. Fusion technology is still in the laboratory stages with much still to be learned. Nuclear fusion facilities are very costly as compared to fission plants.

2. Fission reactions use very heavy elements for nuclear fuel, such as uranium (U-235) and plutonium (Pu-239). Both elements very rare and expensive. Fusion reactions use very light elements for nuclear fuel such as deuterium (D), an isotope of hydrogen which is not so rare or costly. Deuterium is known as "heavy hydrogen," and is found in great abundance in water.

3. Fission fuel is in the form of solid pellets while fusion fuel is, in many cases, a totally ionized gas (plasma).

4. Fission reactions generate large amounts of radiation and radioactive waste products while fusion reactions produce little or no radiation or radioactive waste.

5. Fission reactions are relatively easy to start and control at relatively low temperatures. Conversely, fusion reactions occur at extremely high temperatures and are very difficult to start and control. The reason for this is because it is necessary to force nuclei together. The nuclei contain protons which are positively charged. As a result, the nuclei are positively charged and want to repel one another. To overcome this problem, it is necessary to add energy to the nuclei by superheating the gaseous fuel to above 100,000,000° C. At this temperature, a chain reaction of nuclear collisions will occur. Because of this high temperature requirement, fusion is referred to as a *thermonuclear reaction*. With each collision and subsequent fusion, tremendous amounts of energy in the form of heat are released. As the chain reaction continues, the very high temperature is sustained by the reaction itself. Naturally, there is the very serious problem of being able to heat a gas to such a high temperature to begin with, then to be able to contain the reaction. There are no known materials that can survive 100,000,000° C. Scientists are using various forms of magnetic containment to hold the plasma together. Strong magnetic fields surround the superheated plasma and force it together, at the same time preventing it from

coming in contact with the walls of the plasma chamber. A more thorough explanation of nuclear fusion technology is beyond the scope or intent of this text. You may wish to investigate further by visiting a local library, writing to the U.S. Department of Energy, or researching the topic on the Web.

In spite of the very high cost of facilities and great difficulty of control and containment of the fusion reaction, scientists the world over are continuing their quest to harness this almost utopic source of energy. Low fuel costs, high energy output, and lack of radioactive wastes, make the utilization of nuclear fusion a worthy goal. The conquest of nuclear fusion is one of mankind's greatest frontiers.

7-5. REVIEW QUESTIONS

1. What are the two types of LWRs?
2. Briefly explain how nuclear energy is used to generate electricity.
3. What is nuclear fission?
4. Which are most commonly used, LWRs or breeder reactors?
5. What fuel does a breeder reactor generally use?
6. If the nuclear chain reaction is steadily increasing in the reactor, the reaction is said to be what?
7. Which element is not found in nature: U-235, Pu-239, or D?
8. What is nuclear fusion?
9. List two advantages and two disadvantages of fusion as compared to fission.

Check your answers in the *Answers to Questions* appendix at the back of this book.

7-6. WORLD-WIDE WEB SITES

The Author's Home Page

http://www.castlegate.net/personals/hazen/

Relating to Nuclear Energy

California Energy Commission
    http://www.energy.ca.gov/

Fusion Energy
   http://FusionEd.gat.com/

Nuclear Energy
   http://www.phoenix.net/~nuclear/univ.html

Nuclear Information WWW Server
   http://nuke.westlab.com/

Nuclear Listings
   http://www.energy.ca.gov/energy/earthtext/other.html#NUCLEAR

Oak Ridge National Laboratory
   http://www.ornl.gov

Stanford Nuclear Energy FAQs
   http://steam.stanford.edu/jmc/progress/nuclear-faq.html

Todd's Atomic Homepage
   http://neutrino.nuc.berkeley.edu/neutronics/todd.html

U.S. Nuclear Regulatory Commission
   http://www.nrc.gov/

8

ENERGY STORAGE

The utilization of alternative energy sources has presented three main challenges to mankind: energy collection, energy storage, and the conversion of energy into a usable form. Perhaps the most significant of these is the problem of energy storage. We are well able to collect energy from the sea, the wind, the earth, the sun, and even the atom itself; but we are realizing more and more the importance of storing surplus energy for times of great demand. In modern times, it has become far too costly to build commercial electricity generating plants capable of meeting the peak load demands of today and tomorrow. Utility companies are finding it far more cost effective to build generating plants that have energy storage capabilities. When the electrical demand is low, surplus electrical energy can be converted and stored as another form of energy. During peak load times, the stored energy can then be transformed back into electrical energy to meet the demand. Whether it be a utility company or a domestic energy system, the problem of varying load or demand is a real one, and is economically worth solving. In this chapter, we will explore some of the more common means of solving this problem.

8-1. HYDRO ENERGY STORAGE

Without any doubt, the most common large-scale energy storage method is hydro energy storage. Utility companies the world over have employed this method for decades. The principle of hydro energy storage is a simple one. During low demand periods, water is pumped from a lower reservoir or holding basin to a much higher reservoir. During peak load periods, the water in the higher basin is released to the lower basin through a water turbine electrical generating system. In many cases, the water turbine/generator is also a motor/pump, and will operate equally well in either mode.

8-2. CHEMICAL ENERGY STORAGE

Common, But Not So Common

The next most common method of energy storage is that of chemical energy storage. Batteries are used to convert electrical energy into reactive chemicals. When electrical energy is demanded from the battery, a chemical reaction takes place which generates a large supply of free electrons at the anode terminal (negative) of the battery, thus producing a difference of potential and subsequent current flow in an external circuit. The chemicals involved in the reaction break down into non-reactive chemicals as current is generated and the battery discharges. When the battery is charged and energy is stored, the charging current reverses the chemical reaction, restoring the original reactive chemicals. While chemical energy storage is the next most common means of storing energy, it is not common to high-power utility applications.

Problems With Chemical Energy Storage

The basic problem here is the fact that the batteries are very limited in life or number of charge/discharge cycles, making their cost very high. As a result, utility companies have not made extensive use of chemical energy storage. This is not to say that the utility companies are not interested in batteries. It is not possible, topographically, to utilize hydro-energy storage everywhere on earth. Thus, there is great incentive for utility companies to find other means of energy storage independent of topography. For this reason, much research has gone into new battery designs. The traditional lead-acid battery is not adequate for long life and high-energy storage, as needed by utility companies.

Progress Toward Better Batteries

The Energy Research and Development Administration (ERDA) and the utility industry's Electric Power Research Institute (EPRI) have been very much involved in advancing chemical energy storage technology in the United States. As an example, both organizations have sponsored an experimental battery storage facility in cooperation with the Public Service and Gas Company of Somerset County, NJ. This facility has been named the Battery Energy Storage Test (BEST) facility. While most of the facility's batteries are advanced lead-acid types, the BEST facility is intended as a test bed for new technology batteries that are lighter, smaller, and have much longer cycle lives than comparable lead-acid batteries. Late in 1983, an advanced, high-capacity (500 kWh) zinc-chlorine battery was installed at the BEST facility. This battery has a liquid chlorine electrolyte that is continually circulated between the electrodes in each cell. Other exotic batteries, such as zinc-bromine and sodium-sulfur batteries, are also being tested in a continuing effort to obsolete the old lead-acid designs once and for all. According to ERDA, the new batteries should be able to store five to eight times as much electricity per pound as the lead-acid variety, while occupying one-fourth to one-third the space to produce the same output at a much lower cost. As a result of this and other research, utility companies the world over will soon be taking advantage of this new chemical energy storage technology.

Facts Relating to Lead-Acid and Other Batteries

1. The positive terminal of a battery is correctly called the *cathode*. Like a diode, the cathode is the terminal at which electrons enter. Electrons returning from a load enter the battery via the positive terminal. The negative terminal of a battery is called the *anode*. Like a diode, the anode is the terminal at which electrons exit. You can confirm this by contacting any battery manufacturer.

2. The capacity of a battery is measured in ampere-hours, which is amperes times time. For example, a 110Ah (ampere-hour) battery can supply a current of 1A for 110 hours, or 2A for 55 hours. However, as the load (current drain) increases, the capacity decreases. The 110Ah battery just discussed might be derated to only 80Ah if the load is, say, 30A. Manufacturers provide charts that help you determine a battery's capacity under different loads (current drains).

3. Capacity can also be expressed in watt-hours (Wh). For example, a 12V battery with a 110Ah capacity can also be said to have a 1320Wh capacity. This is because power in watts is equal to voltage times current ($P = V \times A$). Thus, 12V x 110Ah = 1320Wh.

4. The capacity of a battery is affected by temperature. Capacity ratings are established at a temperature near room temperature, usually 25° Celsius. As temperature decreases, so does capacity. At 0° Celsius, the capacity of some lead-acid batteries can be reduced by as much as 50%. This depends on the actual battery.

5. There are six cells in a 12V battery. If it is a lead-acid battery, each cell produces a little more than 2.1V. Thus, a fully-charged, unloaded lead-acid battery produces about 12.7V. A deeply discharged lead-acid battery with only a 10% charge remaining will have an unloaded terminal voltage of about 11.89V. See *Figure 8-1*.

6. A heavily-loaded yet fully-charged lead-acid battery will show a terminal voltage of about 8.5V. "Heavily-loaded" means 100A or more.

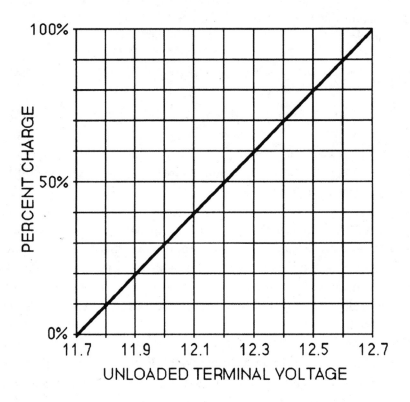

Figure 8-1. *Percent charge vs. unloaded terminal voltage for a typical 12V lead-acid deep-cycle battery.*

7. Lead-acid batteries should never be left uncharged or undercharged. This will cause sulfating of the positive lead plates. Sulfating causes permanent damage that increases the internal resistance of the battery. That means that under heavy loads, the chemical reaction inside the battery will be hindered by the sulfate, and a lesser quantity of electrons will be produced. This causes terminal voltage to sag under load even if the unloaded terminal voltage looks normal.

8. More energy is needed to charge a battery than can be drawn from it. For example, 13.8 to 15 volts is used to charge a lead-acid battery in which the maximum terminal voltage without the charger connected is only about 12.7V. Also, it takes more ampere-hours of charging than can be used during discharge.

9. Most automobile-type lead-acid batteries are shallow-discharge types designed to provide a large amount of current for a short time. These should never be repeatedly discharged more than 20% of their rated capacity.

10. Deep-cycle or deep-discharge batteries are designed to survive repeated deep discharges of as much as 80% (20% remaining) of the rated capacity. However, even these batteries will have their lives extended if the discharges are not deep. Such batteries are used for electric vehicles and renewable energy storage.

11. Batteries in continuous service, such as for a renewable energy storage system, should have a low-voltage load disconnect circuit. This circuit is set to a minimum terminal voltage at which the load is disconnected to prevent damage to the battery system.

12. Overcharging lead-acid batteries produces hydrogen gas, which is highly explosive. Unsealed and capped lead-acid batteries are especially dangerous since the hydrogen can escape and easily be ignited by any spark, such as connecting leads to the terminals or removing a charger clip. Always charge batteries in a well-ventilated area. Battery compartments must be well ventilated.

13. Charging a lead-acid battery should be done under control of a special regulator to ensure long life. Ideally, there should be three stages to the charging process:

    A. The bulk-charge phase when the battery is charged at a relatively high constant rate, not to exceed the manufacturer's recommendation for the particular battery (usually less than Ah/5, assuming a 5A charging current).

    B. The float phase (a.k.a. absorption phase) during which the applied charge voltage is held constant at a recommended level (usually 14 to 14.5V) until charge current decreases to a few amperes.

    C. The trickle phase that eventually brings the battery to full charge and keeps it there (usually 13 to 13.5V).

14. If the battery is unsealed and has caps, you can use a good hydrometer to check the specific gravity of each cell to determine percent of charge. At 77° F (25° C), 100% = 1.265, 75% = 1.239, 50% = 1.200, 25% = 1.170, and 0% = 1.110.

15. Do not add water to a deeply discharged battery. When the battery is charged, the electrolyte is rebuilt and will overflow the caps.

16. Keep the terminals and outer casing of the battery clean and dry. This will prevent excessive terminal corrosion and additional battery drain.

8-3. ELECTROLYSIS ENERGY CONVERSION & STORAGE

Electrolysis and Energy Storage

Electrolysis is a process by which a DC electric current is passed through a conductive liquid (electrolyte), causing a chemical reaction to take place in which certain elements are removed from a compound or solution. So, what does electrolysis have to do with energy storage? The electrolysis of water has been seriously considered, although not widely used, as a means of converting electrical energy into a very useful fuel (hydrogen). In the water electrolysis process, electrical energy is used to separate out hydrogen and oxygen. In this way, the electrical energy is stored as a very volatile fuel, hydrogen, contained under high pressure as a liquid. The liquid hydrogen can be stockpiled or transported where needed to power combustion engines used to drive generators.

The Electrolysis Process

A simple electrolysis system is composed of a DC electrical source, two electrodes (usually graphite), a container for the electrolyte (salt water), and a gas-collecting apparatus. As direct current passes between the electrodes through the electrolyte, hydrogen is separated out of the water molecules and collected at the cathode (negative electrode). At the same time, liberated oxygen atoms collect at the anode. The hydrogen gas is collected by a special apparatus and compressed into a storage container. The high pressure forces the hydrogen into a liquid state.

A Practical Application

Since salt water is needed for electrolysis, this method of energy conversion/storage has been viewed as compatible with OTEC systems. OTEC systems must operate in optimum waters, usually at great distances from shore. This makes cabling electricity to shore difficult due to potentially heavy electrical losses and high cost. A possible alternative would be to manufacture hydrogen through electrolysis at the OTEC platform and ship it to land to be used as a fuel.

8-4. COMPRESSED AIR ENERGY STORAGE

Energy can be stored in compressed air. An alternative energy source can be used to power a high-pressure pump to store a massive volume of compressed air. The compressed air can then be used to drive special air turbines which, in turn, drive electrical generators. To be practical for commercial applications, huge storage chambers would be required that can withstand 500 to 2000 pounds per square inch of pressure. Scientists have considered using deep underground, nonporous caverns for compressed air containment. In fact, this method is being used at a 290 MW compressed air plant in Huntorf, Germany, and a 25 MW plant in Sesta, Italy. As a rule, this method of energy storage is not commonly employed due to the near-ideal geological requirements or prohibitive cost of manmade containers.

8-5. FLYWHEEL ENERGY STORAGE

Flywheel energy storage is often referred to as *inertial energy storage*, since energy is stored in the inertia of a massive flywheel. Flywheels used for energy storage are heavy discs that absorb energy as they spin at tens of thousands of RPM. The initial source of energy may be an alternative energy source or the breaking action of a vehicle. Energy is removed from the spinning flywheel through a special hydraulic transmission as needed. The flywheel must be perfectly balanced and made of a material that will not disintegrate at high RPM. To lessen friction losses, the flywheel is housed in an evacuated chamber (air removed), and special magnetic bearings are often used.

The flywheel has many attractive advantages over other means of energy storage. First of all is the fact that the flywheel can be charged and discharged an almost limitless number of times, unlike batteries. Maintenance is very minimal. There are no potentially dangerous by-products, such as hydrogen. The flywheel can be charged and discharged very quickly, a tremendous advantage over chemical storage systems. Finally, initial cost and operating costs are far less than other storage systems. The greatest challenges presented to flywheel researchers have been to minimize friction losses, optimize drive and control systems, and to design high-speed flywheels that will not disintegrate at higher speeds.

The U.S. Department of Energy (DOE) has been sponsoring flywheel research for applications in many important areas. The following list of possible applications will help you realize the significance of flywheel technology:

1. *Energy conservation in vehicles.* Flywheels have already been used in trains, subways, buses, and automobiles to recover energy that normally would be lost as heat in braking systems. When bringing the vehicle to a stop, the flywheel is engaged and acts as an inertial drag. The energy required to overcome the inertia of the vehicle, in bringing it to a stop, is transferred to the flywheel, causing it to spin. As the vehicle comes to a rest, the flywheel spins at a very high RPM. The energy stored in the spinning flywheel can then be used to accelerate the vehicle. This conserves large amounts of fuel in combustion engine vehicles and conserves battery charge in electric vehicles.

2. *Flywheel-powered vehicles.* The DOE is also doing research in the area of totally fly-wheel-powered vehicles. In this type of vehicle, the only means of propulsion would be the flywheel. The flywheel is charged (spun at high speeds) by a fixed energy source, then is discharged as the vehicle is operated. This is very similar to the small toy cars that have friction or traction motors. The toy car is pushed several times by hand to accelerate a small flywheel as the gears scream loudly. When the toy car is released, it races across a surface, discharging the flywheel.

3. *Wind turbine energy storage.* Another possible application for the flywheel is energy storage for wind turbine systems. The many advantages of the flywheel over battery storage has attracted the attention of many researchers. This kind of energy storage can greatly reduce wind system cost, and overcome the problems of wind availability and constancy.

4. *Domestic energy storage.* Using the flywheel to store energy in the home is also a very real possibility. The flywheel can be charged from an electric motor during low electric demand times. Thus, energy can be collected and stored at times when the electric rates are lower. The flywheel can then be discharged to power an electric generator during expensive peak load times. This will help to level out the load for the utility company and save money for the consumer.

5. *Utility company energy storage.* Large flywheels may be used by utility companies in the same manner as described for domestic use. The greatest problem facing the utility com-panies is having equipment capable of meeting peak load demands. Usually, more equip-ment must be brought on line during peak load times to satisfy the demand. A flywheel system would allow the main generating system to operate at a near-constant load at all times with the flywheel absorbing load variations. At the present time, however, the flywheel technology needed for such massive energy storage is not fully developed.

In this chapter, you have explored some of the many types of energy storage systems currently in use or being considered by researchers. It is fair to say that we have only scratched the surface in this investigation. Further information can be found at your local library, the U.S. Department of Energy, and/or by doing a search on the Web.

8-6. REVIEW QUESTIONS

1. What is the most common form of energy storage used by utility companies?
2. Water electrolysis to produce hydrogen is best suited for what type of alternative energy source?
3. How can compressed air be used to store energy?
4. What are the advantages of the flywheel over chemical energy storage?

Check your answers in the *Answers to Questions* appendix in the back of this book.

8-7. WORLD-WIDE WEB SITES

The Author's Home Page

http://www.castlegate.net/personals/hazen/

Relating to Energy Storage

California Energy Commission
http://www.energy.ca.gov/energy/html/directory.html

9

THE U.S. DEPARTMENT OF ENERGY

In this chapter, we will explore the United States Department of Energy (DOE) and its national laboratories. Since August 4, 1977, when President Carter signed the Department of Energy Organization Act, the DOE has played a vital role in the nation's research, technology and industrial development. Much of the nation's funding for research and technological advances has flowed through the DOE. The DOE itself has been a source of energy for our nation. The information contained in this chapter is provided by the DOE and its national laboratories. (DOE web site, http://www.doe.gov/)

9-1. THE DEPARTMENT'S MISSION (DOE's Mission)

The Department of Energy, in partnership with our customers, is entrusted to contribute to the welfare of the nation by providing the technical information and scientific and educational foundation for technology, policy, and institutional leadership necessary to achieve efficiency in energy use, diversity in energy sources, a more productive and competitive economy, improved environmental quality, and a secure national defense.

9-2. THE DEPARTMENT'S CORE VALUES (DOE's Core Values)

1. We Are Customer-Oriented.
   A. Our decisions and actions are responsive to the customer's needs.
   B. We foster a participatory manner of doing business where the opinions and input of diverse stakeholders are sought and considered early in the decision-making process.
   C. Programs and solutions to major issues are developed in a proactive way with our customers and stakeholders.

2. People Are Our Most Important Resource.
   A. We value the needs of individuals.
   B. We are committed to improving the knowledge, skills, and abilities of our employees by providing opportunities for professional development and achievement.
   C. We are committed to providing a safe and healthy workplace for all our employees.
   D. We value the richness, experience, and imaginative ideas contributed by a diverse work force.
   E. We share credit with all contributors.
   F. We value listening as an essential tool in learning from others.
   G. Employees are forthright in sharing their experiences so that we can learn from each other.

3. Creativity and Innovation are Valued.
   A. We are committed to a flexible operating environment that facilitates the pursuit of new technologies, processes, programmatic approaches, and ideas that challenge the status quo.
   B. We seek out, nurture, and reward innovation in daily activities, ranging from the routine to the complex.
   C. Employees are empowered to pursue creative solutions.
   D. Resourcefulness, efficiency, and effectiveness are recognized and rewarded.
   E. Adaptable, entrepreneurial approaches that can respond quickly to the rapidly changing world business and political environment are essential.

4. We Are Committed to Excellence.
   A. Quality and continuous improvement are essential to our success.
   B. We are committed to excellence in everything we do.
   C. Through scientific excellence, we are committed to enhancing the nation's ability to compete globally.

5. *DOE Works as a Team and Advocates Teamwork.*
   A. We reinforce the notion of a common or greater departmental good and encourage interdepartmental teamwork to achieve this goal.
   B. We value teamwork, participation, and pursuit of win/win solutions as essential elements of our operating style.
   C. We work as a team with other federal agencies, government organizations, and external stakeholders in pursuing broader national goals.
   D. We recognize the needs of others for information and communicate knowledge and information in an open and candid manner.

6. *We Respect the Environment.*
   A. We recognize our leadership role and responsibility to improve the quality of the environment for future generations.
   B. We recognize the importance of the environmental impacts of our operations and develop and employ processes and technologies to reduce or eliminate waste production and pollution in these operations.

7. *Leadership, Empowerment, and Accountability Are Essential.*
   A. We are visionaries in our everyday activities.
   B. Leaders trust and support individuals to make informed decisions about the processes that they own.
   C. We are effective stewards of the taxpayer's interest.
   D. Our actions are result-oriented.

8. *We Pursue the Highest Standards of Ethical Behavior.*
   A. We maintain a personal commitment to professionalism and integrity.
   B. We assure conformance with applicable laws, regulations, and responsible business practices.
   C. We keep our commitments.
   D. We are objective and fair.

9-3. DEPARTMENT OF ENERGY CHRONOLOGY

(Source: THE UNITED STATES DEPARTMENT OF ENERGY, 1977-1994, A Summary History, ENERGY HISTORY SERIES, by Terence R. Fehner and Jack M. Holl, History Division, Executive Secretariat, Human Resources and Administration, Department of Energy.)

June 29, 1973: President Nixon establishes the Energy Policy Office.

October 6, 1973: The Yom Kippur War breaks out in the Mideast.

October 17, 1973: The Organization of Arab Petroleum Exporting Countries declares an oil embargo.

November 7, 1973: President Nixon launches Project Independence.

December 4, 1973: The Federal Energy Office replaces the Energy Policy Office. William Simon is named administrator.

May 7, 1974: President Nixon signs the Federal Administration Act of 1974. The Federal Energy Administration replaces the Federal Energy Office.

August 9, 1974: Gerald R. Ford becomes president.

October 11, 1974: President Ford signs the Energy Reorganization Act of 1974. The Atomic Energy Commission is abolished. The Energy Research and Development Administration, Nuclear Regulatory Commission, and Energy Resources Council are established.

November 25, 1974: President Ford appoints Frank Zarb as administrator, Federal Energy Administration.

January 19, 1975: The Energy Research and Development Administration is activated. President Ford appoints Robert C. Seamans, Jr., as administrator.

December 22, 1975: President Ford signs the Energy Policy and Conservation Act, extending oil price controls into 1979, mandating automobile fuel economy standards, and authorizing creation of a strategic petroleum reserve.

January 20, 1977: Jimmy Carter is inaugurated president.

February 2, 1977: President Carter signs the Emergency Natural Gas Act of 1977.

February 7, 1977: John F. O'Leary is named administrator, Federal Energy Administration.

April 18, 1977: President Carter announces National Energy Plan in his first major energy speech.

August 4, 1977: President Carter signs the Department of Energy Organization Act. The Federal Energy Administration and Energy Research and Development Administration are abolished.

August 5, 1977: James R. Schlesinger is sworn in as first Secretary of Energy.

October 1, 1977: The Department of Energy is activated.

November 9, 1978: President Carter signs the National Energy Act, which includes the National Energy Conservation Policy Act, the Powerplant and Industrial Fuel Use Act, the Public Utilities Regulatory Policy Act, the Energy Tax Act, and the Natural Gas Policy Act.

January 16, 1979: The Shah flees Iran.

March 28, 1979: An accident occurs at the Three Mile Island nuclear power plant.

April 5, 1979: President Carter, responding to growing energy shortages, announces gradual decontrol of oil prices and proposes windfall profits tax.

June 20, 1979: President Carter announces program to increase the nation's use of solar energy, including a solar development bank and increased funds for solar energy research and development.

July 10, 1979: President Carter proclaims a national energy supply shortage and establishes temperature restrictions in nonresidential buildings.

July 15, 1979: President Carter declares energy to be the immediate test of ability to unite the nation, and proposes an $88 billion, decade-long effort to enhance production of synthetic fuels from coal and shale oil reserves.

August 24, 1979: Charles W. Duncan, Jr., is sworn in as Secretary of Energy.

October 1, 1979: Secretary Duncan announces the reorganization of the Department of Energy to manage programs by technologies or fuels.

June 30, 1980: President Carter signs the Energy Security Act, consisting of six major acts: U.S. Synthetic Fuels Corporation Act, Biomass Energy and Alcohol Fuels Act, Renewable Energy Resources Act, Solar Energy and Energy Conservation Act and Solar Energy and Energy Conservation Bank Act, Geothermal Energy Act, and Ocean Thermal Energy Conversion Act.

January 20, 1981: Ronald Reagan is inaugurated president.

January 23, 1981: James B. Edwards is sworn in as Secretary of Energy.

January 28, 1981: President Reagan signs Executive Order 12287, which provides for the decontrol of crude oil and refined petroleum products.

February 18, 1981: President Reagan presents America's New Beginning: A Program for Economic Recovery to Congress.

February 25, 1981: Secretary Edwards announces a major reorganization of the Department of Energy to improve management and increase emphasis on research, development, and production.

February 25, 1981: Secretary Edwards creates the Energy Policy Task Force.

July 17, 1981: The Department of Energy releases third national energy policy plan, Securing America's Energy Future: The National Energy Policy Plan.

October 8, 1981: The Reagan Administration announces a nuclear energy policy that anticipates the establishment of a facility for the storage of high-level radioactive waste and lifts the ban on commercial reprocessing of nuclear fuel.

February 1982: Secretary Edwards sends to Congress the Sunset Review, a comprehensive review of each departmental program required by the Department of Energy Organization Act of 1977.

April 5, 1982: Secretary Edwards announces placement of the 250-millionth barrel of oil in the Strategic Petroleum Reserve.

May 24, 1982: President Reagan proposes legislation transferring most responsibilities of the Department of Energy to the Department of Commerce.

November 11, 1982: Donald Paul Hodel is sworn in as Secretary of Energy.

January 7, 1983: President Reagan signs the Nuclear Waste Policy Act of 1982, the nation's first comprehensive nuclear waste legislation.

July 21, 1983: President Reagan endorses the Alternative Financing Plan and restates his support for the Clinch River Breeder Reactor Plant Project.

October 4, 1983: President Reagan presents the fourth National Energy Policy Plan, with a goal of fostering an adequate supply of energy at reasonable costs to Congress.

October 7, 1983: The Department of Energy establishes the Office of Civilian Radioactive Waste Management.

October 26, 1983: The Senate refuses to continue funding the Clinch River Breeder Reactor, effectively terminating the project.

January 20, 1984: Secretary Hodel announces a reorganization plan to improve the management of programs critical to the nation's energy security.

May 8, 1984: Secretary Hodel gives the Nuclear Power Assembly his assessment of the state of the nuclear power industry, and urges action on the administration's proposed nuclear plant licensing reform bill.

October 25, 1984: The National Coal Council is established to advise both government and industry on ways to improve cooperation in areas of coal research, production, transportation, marketing, and use.

January 3, 1985: Secretary Hodel transmits the natural gas report to Congress, urging comprehensive deregulation.

February 7, 1985: John S. Herrington is sworn in as Secretary of Energy.

September 18, 1985: Secretary Herrington consolidates the department's environment, safety, and health activities under a newly created assistant secretary.

November 13, 1985: Secretary Herrington outlines his five-point strategy to help revitalize the nation's nuclear industry in a speech before a joint meeting of the Atomic Industrial Forum and the American Nuclear Society.

February 24, 1986: Technical safety appraisals begin for more than fifty Department of Energy facilities in eleven states.

March 26, 1986: The fifth National Energy Policy Plan, outlining the continued goal of an adequate supply of energy available at a reasonable cost, is submitted to Congress.

April 3, 1986: Successful reactor safety tests are conducted at the Experimental Breeder Reactor in Idaho.

April 10, 1986: Secretary Herrington asks Congress to open access to interstate natural gas pipelines and lift all remaining controls on natural gas prices.

April 26, 1986: A Soviet nuclear reactor accident occurs at Chernobyl.

May 14, 1986: Secretary Herrington requests the NAS/NAE to make an independent safety assessment of the Department of Energy's eleven major production and research reactors.

May 28, 1986: Three candidate sites are selected for the first high-level nuclear waste repository.

September 24-29, 1986: Secretary Herrington leads the U.S. delegation to a special session of the IAEA General Conference in Vienna, Austria, to discuss measures to strengthen international cooperation in nuclear safety, and radiological protection in the aftermath of Chernobyl.

January 30, 1987: Secretary Herrington announces President Reagan's approval of construction of the superconducting super collider, the world's largest and most advanced particle accelerator.

February 18, 1987: The Department of Energy report, *America's Clean Coal Commitment*, catalogs thirty-seven projects underway or planned for clean coal demonstration facilities.

March 17, 1987: The Department of Energy report, *Energy Security*, outlines the nation's increasing dependence on foreign oil.

April 1, 1987: The Department of Energy issues an invitation for site proposals for the superconducting super collider.

July 28-29, 1987: President Reagan announces an eleven-point superconductivity initiative at the Federal Conference on Commercial Applications of Superconductivity, sponsored jointly by the Department of Energy and the White House Office of Science and Technology Policy.

October 1, 1987: The Department of Energy celebrates its tenth anniversary.

December 22, 1987: Congress approves an amendment designating Yucca Mountain, Nevada, as the only site to be considered for the high-level nuclear waste repository.

January 19, 1988: Secretary Herrington announces seven best qualified sites for the superconducting super collider, located in Arizona, Colorado, Illinois, Michigan, North Carolina, Tennessee, and Texas.

August 3, 1988: Secretary Herrington announces decision to build two new production reactors: a heavy water reactor at the Savannah River Plant, and a modular, high-temperature, gas-cooled reactor to be located at the Idaho National Engineering Laboratory.

August 23, 1988: President Reagan signs an omnibus trade bill that repeals the Windfall Profits tax.

November 10, 1988: Secretary Herrington designates the Texas site for the superconducting super collider.

January 12, 1989: The White House releases the 2010 Report, projecting requirements for maintaining and modernizing the nuclear weapons production complex through the year 2010.

January 20, 1989: George Bush is inaugurated president.

March 9, 1989: James D. Watkins is sworn in as Secretary of Energy.

March 23, 1989: Scientists at the University of Utah announce the discovery of cold fusion, drawing immediate worldwide attention.

June 6, 1989: The Justice Department announces an investigation into possible violations of federal environmental laws at Rocky Flats.

June 27, 1989: Secretary Watkins announces a plan to strengthen environmental protection and waste management activities at the department's production, research and testing facilities.

July 6, 1989: Nevada Governor Robert Miller signs a bill declaring storage of high-level radioactive waste in the state to be illegal.

July 26, 1989: President Bush directs the Department of Energy to develop a comprehensive national energy policy plan.

August 1, 1989: Secretary Watkins announces the completion of the five-year cleanup plan to characterize and prioritize waste cleanups at departmental sites.

September 29, 1989: Secretary Watkins establishes the Modernization Review Committee to review the assumptions and recommendations of the 2010 Report.

November 9, 1989: Secretary Watkins establishes the Office of Environmental Restoration and Waste Management within the department.

November 28, 1989: The Department of Energy announces a new high-level waste management plan, and asks the Justice Department to file suit to obtain necessary permits for the Yucca Mountain repository.

August 2, 1990: Iraq invades and seizes Kuwait, creating a major international crisis.

August 15, 1990: Secretary Watkins announces plans to increase oil production and decrease consumption to counter Iraqi-Kuwaiti oil losses.

November 21, 1990: President Bush declares the end of the Cold War as relations ease with the Soviet Union.

December 21, 1990: Secretary Watkins presents the National Energy Strategy to President Bush.

January 11, 1991: The IEA Governing Board agrees to a contingency plan combining a stockdraw with demand restraint measures to prevent sharp oil price increases in the event of war.

January 16-17, 1991: United Nations coalition forces launch Operation Desert Storm when Saddam Hussein refuses to withdraw from Kuwait.

January 28, 1991: The Department of Energy obtains an administrative land withdrawal from the Department of Interior, giving the DOE full control over the Waste Isolation Pilot Plant (WIPP).

February 7, 1991: The Complex Reconfiguration Committee, formerly the Modernization Review Board, releases its recommendations for a reconfigured weapons complex, Complex-21.

February 20, 1991: President Bush presents the Department of Energy's National Energy Strategy to Congress and the American people.

March 4, 1991: Secretary Watkins transmits the Bush Administration's energy bill to the House and Senate.

July 31, 1991: President Bush signs the Strategic Arms Reduction Treaty (START), which will reduce nuclear weapon stockpiles to 6,000 accountable warheads.

September 27, 1991: President Bush announces additional unilateral cuts in the nuclear weapon arsenal.

January 31, 1992: A federal judge rules administrative land withdrawal for WIPP invalid.

May 10, 1992: Secretary Watkins testifies before the Senate Armed Services Committee that for the first time since 1945, the United States is not building any nuclear weapons.

June 1992: Representatives from many nations attend the Earth Summit in Rio de Janeiro.

September 1992: Congress votes to impose a nine-month moratorium on nuclear weapons testing.

October 24, 1992: President Bush signs the Energy Policy Act of 1992, which assists the implementation of the National Energy Strategy.

October 30, 1992: President Bush signs the WIPP Land Withdrawal Act.

November 3, 1992: William "Bill" Clinton is elected president.

January 22, 1993: Hazel R. O'Leary is sworn in as Secretary of Energy.

February 17, 1993: President Clinton reveals his economic recovery plan in his State of the Union message.

April 2, 1993: Secretary O'Leary reorganizes the Department of Energy by missions: energy, weapons and waste cleanup, and science and technology.

April 21, 1993: President Clinton announces that the United States will stabilize greenhouse gas emissions at 1990 levels by the year 2000.

July 3, 1993: President Clinton extends the nuclear weapons testing moratorium for at least fifteen months.

October 1993: Congress votes to terminate the superconducting super collider.

October 19, 1993: President Clinton and Vice President Gore unveil the Climate Change Action Plan, emphasizing voluntary measures to stabilize greenhouse gas emissions.

December 7, 1993: Secretary O'Leary announces her openness initiative.

January 3, 1994: The Human Radiation Interagency Working Group, tasked with coordinating the search for human experimentation records, holds its first meeting.

April 1994: Secretary O'Leary releases her strategic plan for the Department of Energy.

9-4. FUELING A COMPETITIVE ECONOMY: DOE'S STRATEGIC PLAN

(Source: THE UNITED STATES DEPARTMENT OF ENERGY, 1977-1994, A Summary History, ENERGY HISTORY SERIES, by Terence R. Fehner and Jack M. Holl, History Division, Executive Secretariat, Human Resources and Administration, Department of Energy.)

Culminating months of effort, Secretary O'Leary in April, 1994, released the Department of Energy's first comprehensive strategic plan. O'Leary noted that the end of the Cold War and the election of President Clinton had engaged a new national agenda. Beginning with the summer of 1993 empowerment summit at the Motorola-Miliken Quality Institute, and through the process of a total quality management learning experience, the strategic planning process envisioned a massive reshaping of the department's missions, priorities, and business practices to meet the challenge of the new national agenda. Tinkering around the edges, the strategic plan declared, was not enough. According to O'Leary, the strategic planning process thus produced new and more sharply focused goals: fueling a competitive economy, improving the environment through waste management and pollution prevention, and reducing nuclear danger.

The key to meeting these goals was the effort to define and integrate the business activities of the DOE. The strategic plan identified five core businesses or mission areas.

Five Core Businesses or Mission Areas

1. *Industrial Competitiveness.* To assist President Clinton in achieving his vision of an investment-driven economy capable of creating high-wage jobs, the DOE set as its first priority helping the nation's industry compete in a global economy. This required partnering with industry in research and development to drive products into the marketplace and cut costs through greater resource efficiency and pollution prevention.

2. *Energy Resources.* Convinced that economic growth, energy security, and environmental preservation were not irreconcilable goals, the DOE reiterated support for sustainable energy technologies, emphasizing energy efficiency, renewable resources, and the economic and clean use of fossil fuels. Favoring technological command and control solutions, the strategic plan promoted diversity and flexibility in energy sources and stressed the need for economic and regional equity for all Americans.

3. *National Security.* For nearly five decades, the defense programs of the DOE and its predecessor agencies focused on the threat of nuclear conflict. The new danger, according to the strategic plan, was the proliferation of nuclear weapons and materials by rogue states and terrorist groups. The DOE's redirected national security mission, therefore, concentrated on nonproliferation, safe dismantlement of nuclear weapons, and maintenance of the stockpile without nuclear testing.

4. *Environmental Quality.* The strategic plan stated that the DOE's greatest challenge was to eliminate the risks and imminent threats posed by past departmental activities and decisions. Noting the Clinton Administration's commitment to honoring the government's

obligation in addressing nuclear complex cleanup, and high-level nuclear waste from nuclear power plants, the DOE promised to reduce environmental, safety, and health risks while developing technologies and institutions required to solve domestic and global environmental problems.

5. *Science and Technology.* With the nation's industry increasingly shifting from long-term and basic research to short-term product development and improvement, the strategic plan projected the necessity not only to help industry compete effectively in the near-term, but also to meet the needs for long-term research. This required, careful manage ment of the DOE's scientific portfolio in order balance basic and applied research needs. In addition, the DOE hoped to maintain the nation's global technical leadership through long-term, systematic reform of science and mathematics education.

Science and technology were indeed the linchpin uniting the DOE and its various businesses around a common theme. Science and technology, the strategic plan noted, provided the core competencies that enabled all of the DOE's businesses to succeed in their missions. Clearly, as Secretary O'Leary put it, the DOE possessed extraordinary scientific and technical talent and resources. These included 30,000 scientists and engineers, fifty-eight of whom (the strategic plan pointed out) were Nobel Prize winners, employed at nine major multi-program laboratories, ten single-purpose laboratories, eleven smaller special-mission laboratories, and a wide range of special user facilities. Capital value of the laboratories was $30 billion, with annual departmental expenditures of $7 billion for research and development. This represented nearly 10 percent of total federal research and development spending. In essence, the DOE was a scientific and technological agency. As the DOE's new mission statement declared, the DOE contributed to the welfare of the nation by providing technical information, and a scientific and educational foundation.

Finally, for Secretary O'Leary and her strategic planners, the way business was conducted was as crucial as the nature of the business. Embracing continuous quality improvement, the strategic plan identified four critical success factors for the operation of the DOE's businesses:

1. Communicating information and building trust both within the organization and with stakeholders and customers.

2. Focusing on people as the DOE's most important resource by providing employee training, rewarding performance, and promoting work force diversity.

3. Ensuring the safety and health of workers and the public, and protecting and restoring the environment.

4. Managing materials and operations more cost-effectively to give the DOE greater flexibility.

Above all, the DOE needed to be customer oriented. The DOE needed, Secretary O'Leary asserted, the advice and thinking of the broad array of stakeholders and customers.

Whither the Department?

For the Department of Energy, the first year-and-a-half with a new administration and a new secretary was an active one. *Change* was clearly the watchword. As the chart at Secretary O'Leary's initial budget briefing in April, 1993, declared in big, bold letters: *We Changed Our Priorities*. Decades-old functions and activities that descended through the DOE as traditions underwent intense scrutiny to determine if they were still needed and helpful in the new, post-Cold War world. Some were found wanting. Others emerged reformed and revitalized. According to many observers, a greater sense of departmental unity and purpose began to appear.

The strategic planning process was a major step in this direction. The strategic plan envisioned a new Department of Energy with new priorities and a sense of purpose, a new vigilance, and a culture and values that bore no resemblance to the previous organization that grew out of the Cold War. If the long-term shape and scope of the DOE remained as yet uncertain and still evolving, there was obviously no lacking of vision and a sense of the future. While the Department of Energy neither could nor should forget its history and where it came from, there was little doubt that the Department could never return to what it was.

9-5. THE OFFICE OF THE SECRETARY

The Secretary of the Department of Energy, as chief executive officer, provides the overall vision, programmatic leadership, management direction, and administration of the DOE; decides major energy policy and planning issues; acts as the principal spokesperson for the DOE; and assures that effective communication and working relationships with state, local and tribal governments, the president, Congress, other federal departments and agencies, the private sector, and the public are achieved. The secretary is the principal adviser to the president on energy policies, plans, and programs.

In the absence of the secretary, the Deputy Secretary of the DOE acts for the secretary, and assists the secretary in deciding major energy policy and planning issues and in representing the DOE before Congress and the public. The deputy secretary, as chief operating officer, assisted by the under secretary, provides the DOE with daily management, guidance and decision-making, and coordinates the efforts of the energy, weapons/waste cleanup, and science and technology programs to achieve the DOE's goals by delivering quality services to the public.

The deputy secretary has primary oversight responsibility for the DOE's energy efficiency and renewable energy programs; fossil energy; nuclear energy; energy information; civilian radioactive waste management; and the power marketing administrations.

The Under Secretary of the Department of Energy has primary responsibility for the DOE's defense programs; environmental restoration and waste management; intelligence and national security programs; energy research; science education and technical information programs; and laboratory management.

The O'Leary Administration (1993 -)

The O'Leary administration is committed to developing balanced, integrated energy policies which support the nation's economic growth, environmental quality, and educational excellence, today and tomorrow. Secretary Hazel R. O'Leary testified at her confirmation hearing in January, 1993, that under her leadership, the Department of Energy will be open and collegial in its approach to all stakeholders, that it will strive for excellence and efficiency in all its programs, and that its performance will be measured in terms of benefits to the American people. Her priorities include worker health and safety; environmental cleanup and effective waste management; energy efficiency, alternative fuels, and renewable energy technologies; and defense conversion and redirection of the national laboratories to support economic growth and jobs as well as national security.

9-6. DOE BUDGET REQUESTS

Table 9-1 and *Table 9-2* show the DOE's budget requests, for energy research and development, and the total DOE budget during 1980, 1985, 1990, and 1995. Note the decreased budgets for energy research and development since 1980 compared with the total DOE budget since 1980.

Department of Energy Budget Requests
FY 1980, FY 1985, FY 1990, FY 1995
(in millions of dollars)

	1980[1]	1985	1990	1995
Energy Research & Development	3810	2409	2375	3400
Basic Science (includes SSC)	474	746	1169	1337
Conservation Grants	328	252	8	325
Direct Energy Production (includes SPR)	157	1332	1134	1091
Defense	3022	7806	7882	5630
Defense Waste (Environmental Management)	---[2]	---[2]	1145	6521
ES&H and Related Functions	---[2]	---[2]	125	169
Nuclear Waste Repository	---	328	740	533
Regulation & Information	323	114	197	103
Policy, Management & Misc.	308	219	265	285
Adjustments	---	-391	-49	-941[3]
Total	8422	12815	14991	18453

[1]First DOE Budget request as a comprehensive document and not as a combination of requests of predecessor agencies.
[2]No figures available. Amounts subsumed in other categories.
[3]Use of prior year balances and other adjustments.

Table 9-1. *Department of Energy budget requests. Courtesy of the U.S. Department of Energy, FY 1980 Budget to Congress: Budget Highlights (Washington: DOE/CR-004); Department of Energy, FY 1985 Budget Highlights (Washington: DOE/MA-0062/2); Department of Energy, Fiscal Year 1990 Budget Highlights (Washington: DOE/MA-0357); Department of Energy, FY 1995 Budget Highlights (Washington: DOE/CR-0019).*

9-7. MAJOR DOE NATIONAL LABORATORIES

In this section, we will visit some of the major DOE national laboratories. In each case, for the purpose of accuracy, information that is presented was taken from each laboratory's home page residing on the Internet. *Figure 9-1* is a map showing the locations of DOE's Internet servers and national laboratories.

Department of Energy Budget Requests
Energy Research & Development
FY 1980, FY 1985, FY 1990, FY 1995
(in millions of dollars)

Energy Research & Development	1980[1]	1985	1990	1995
Fossil	796	273	165	483
Clean Coal	---	---	325	37
Solar	597	164	71	301
Geothermal	111	27	15	37
Hydroelectric	18	1	---	1
Fusion	364	483	349	373
Nuclear Fission	1037	618	353	248
Environmental	278	228	271	435
Basic Energy	276	480	590	741
Conservation	227	148	88	685
Other	106	40	148	59
Savings from Management Initiatives	---	-53	---	---
Total	3810	2409	2375	3400

[1]First DOE Budget request as a comprehensive document and not as a combination of requests of predecessor agencies.

Table 9-2. *Department of Energy research and development budget requests. Courtesy of the U.S. Department of Energy, FY 1980 Budget to Congress: Budget Highlights (Washington: DOE/CR-004); Department of Energy, FY 1985 Budget Highlights (Washington: DOE/MA-0062/2); Department of Energy, Fiscal Year 1990 Budget Highlights (Washington: DOE/MA-0357); Department of Energy, FY 1995 Budget Highlights (Washington: DOE/CR-0019).*

Ames Laboratory

(U.S. Department of Energy, Ames, Iowa, http://www.ameslab.gov/)

Ames Laboratory is a government-owned U.S. Department of Energy laboratory operated by Iowa State University. Ames Laboratory seeks solutions to energy-related problems through the exploration of chemical, engineering, materials, mathematical and physical sciences. Established in the 1940s with the successful development of the most efficient process to produce high-purity uranium metal for atomic energy, Ames Laboratory now pursues a wide range of material and environmental research, which has strengthened the laboratory's international prominence.

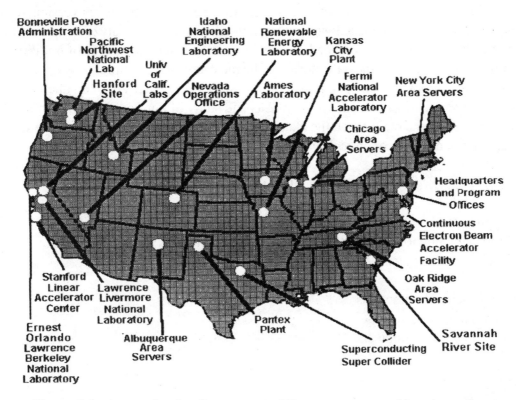

Figure 9-1. *A map showing Department of Energy servers, and locations of national laboratories. Courtesy of the U.S. Department of Energy.*

The Mission of Ames Laboratory

Ames Laboratory effectively focuses diverse fundamental and applied research strengths upon issues of national concern, cultivates tomorrow's research talent, and develops and transfers technologies to improve industrial competitiveness and enhance U.S. economic security. At the forefront of current materials research, high performance computing, and environmental science and management efforts, the laboratory seeks solutions to energy-related problems through the exploration of physics, chemistry, engineering, applied mathematics and materials sciences. All operations are conducted so as to maintain the health and safety of all workers and with a genuine concern for the environment.

Established in the 1940s with the successful development of the most efficient process to produce uranium for the Manhattan Project, Ames Laboratory now pursues much broader priorities than the materials research which has given the lab international credibility. Examples of specific projects include: world-class fundamental photosynthesis studies to ultimately help in the design of synthetic molecules for direct solar energy conversion; development of a remote-controlled analysis system that will acquire and analyze samples from hazardous waste sites at greatly reduced risk and cost; research to break free of traditional programming methods and harness the power of the most advanced computing systems available for scientists unlocking the secrets of revolutionary new materials like superconductors, fullerenes and quasicrystals; and the synthesis and study of nontraditional materials such as organic polymers and organometallic materials to serve as novel semiconductors, processable preceramics and nonlinear optical systems.

Ames Laboratory's research projects fall within twelve program areas:

1. Applied Mathematical Sciences.
2. Environmental Restoration & Waste Management (EM).
3. Environmental Sciences.
4. Fossil Energy.
5. Fundamental Interactions.
6. Materials Chemistry.
7. Metallurgy & Ceramics.
8. Nondestructive Evaluation.
9. Processes & Techniques.
10. Safeguards & Security.
11. Condensed Matter Physics.
12. Advanced Nuclear Systems.

Argonne National Laboratory (ANL)
(DOE's Largest Energy Research Center; http://www.anl.gov/)

Argonne National Laboratory is one of the U.S. Department of Energy's largest research centers, with an annual operating budget of about $490 million supporting more than 200 research projects. Argonne was the nation's first national laboratory. Today, the laboratory has about 4,500 employees, including about 1,775 scientists and engineers, of whom about 800 hold doctorate degrees.

Argonne research falls into four broad categories:

1. The Advanced Photon Source, now under construction, will provide the nation's most brilliant X-ray beams for pioneering research in materials science, a cornerstone of technological competitiveness. Experiments are scheduled to begin in 1996.

2. Energy and environmental science and technology includes research in biology, alternate energy systems, environmental assessments, economic impact assessments, and urban technology development.

3. Engineering research focuses on advanced batteries and fuel cells, and advanced fission reactor systems — including electrochemical treatment of spent DOE fuel for disposal, improved safety of Soviet-designed reactors, and technology for decontaminating and decommissioning aging reactors.

4. Physical research includes materials, science, physics, chemistry, high-energy physics, mathematics and computer science — including high-performance computing and massively parallel computers.

Industrial technology development is an important activity in moving the benefits of Argonne's publicly-funded research to industries to help strengthen the nation's technology base. This activity at Argonne grows out of basic and applied research in materials, transportation, computing, advanced manufacturing, energy, the environment, and other areas. The laboratory maintains an electronically-searchable database of programs and capabilities in which partnerships with industry are invited. Argonne transfers technology to industry through collaborative research and development agreements (CRADAs), licensing, work for others, and "quick-response" mechanisms. Since 1984, Argonne has been the source of more than 30 spin-off companies.

Another important part of Argonne's mission is to design, build and operate national user facilities. These facilities, such as the Advanced Photon Source and the Intense Pulsed Neutron Source, are large, sophisticated research facilities that would be too expensive for a single company or university to build and operate. They are used by scientists from Argonne, industry, colleges and universities and other national laboratories, and often by scientists from other nations.

Argonne's Division of Educational Programs provides a wide range of educational opportunities for faculty and students ranging from leading national universities to local junior high schools. More people attend educational programs at Argonne than at any other DOE national laboratory. Argonne is operated by the University of Chicago for the U.S. Department of Energy.

Fermi National Accelerator Laboratory (FNAL)
(http://fnnews.fnal.gov/)

Fermilab is a high-energy physics laboratory, home of the world's most powerful particle accelerator, the Tevatron. Scientists from across the U.S. and around the world use Fermilab's resources in experiments to explore the most fundamental particles and forces of nature. By any measure — the number of physicists doing experiments, the number of publications reporting new physics results, the level of accelerator performance — Fermilab is where the field of experimental high-energy physics does most of its work in the United States. The laboratory and its accelerator give experimenters a unique capability for gaining new insight into the nature of matter and energy for future technologies, and maintains world leadership in science in support of the Department of Energy's mission of service to the nation.

> *"For the next decade, Fermilab will remain the world's outpost on the energy frontier."*
> — **John Peoples, Jr., Fermilab Director**

Fermilab is a single-program laboratory whose mission is to provide leadership and resources for basic research at the frontiers of elementary particle physics and related disciplines. Fermilab fosters and stimulates science education, transfers to industry technologies developed at the laboratory, and conducts operations with the goal of excellence in health, safety, and the protection of the environment.

The research conducted at Fermilab has helped in major ways to confirm and refine the modern view of the particles and forces in nature. The information provided by the many experiments performed at Fermilab, together with experiments done elsewhere, has been crucial to the great progress that is now taking place in understanding the current state and evolution of the universe. In addition to experimental research, Fermilab also has a strong program in theoretical physics.

Lawrence Livermore National Laboratory (LLNL)
(http://www.llnl.gov/)

Lawrence Livermore National Laboratory (LLNL) is operated by the University of California under a contract with the U.S. Department of Energy.

LLNL Vision and Goals

Today's world requires a fresh assessment of the role of the national laboratories in general, and the role of the defense laboratories and Livermore in particular. The future of the laboratory demands change. The Department of Energy's national laboratories were born out of an urgent need to direct the best that science and technology could offer to the most pressing national issues. The three national defense laboratories were created because it was believed that they could accomplish their vital mission faster, better, and with greater certainty than any combination of private industry and universities. The remarkable success of that initiative remains the foundation of these laboratories and inspires efforts today.

Today, the challenges facing the nation have changed, but the need for the national laboratories remains. Experience has shown that these laboratories are most valuable when:

1. The national interest is at stake.
2. The best science and technology are required.
3. Large and complex research facilities are needed.
4. Expertise in a variety of disciplines must be integrated.
5. The technical risk is high, with the potential of very high rewards.
6. A sustained commitment is needed.
7. The job will go undone if the national laboratories don't do it.

Livermore is changing to meet today's challenges. By matching the laboratory's areas of expertise to pressing national and global challenges, Livermore is focusing on three areas of long-term importance where its contributions are unique and valuable:

1. Global Security: Reducing the nuclear danger.
2. Global Ecology: Harmonizing the economy with the environment.
3. Bioscience: The new frontier.

Livermore's global security program has two major thrusts: to reduce the nuclear danger by ensuring confidence in the safety, security, and performance of the U.S. stockpile; and to prevent and counter nuclear proliferation by applying its expertise in nuclear science and technology. The safe and secure "build-down" of the world's stockpiled weapons will be a continuing responsibility. Livermore will be carrying out these responsibilities in concert with the other two defense laboratories as part of the Department of Energy's integrated plan. The application of its advanced defense technologies will significantly enhance the nation's ability to use non-nuclear means for containing regional conflicts.

Harmonizing the demands of the world's economy with the needs of the environment is a crucial national and global issue. Achieving this balance will require energy sources that are safe and clean as well as manufacturing processes and consumer goods that make wise use of resources and provide for the protection of the environment. Livermore can contribute to all aspects of this challenge—developing energy sources, working with industry to devise advanced manufacturing processes, and developing innovative and cost-effective technologies for environmental management and cleanup.

Bioscience is the new frontier of research. For the first time in history, tools exist to decipher the genetic blueprint (DNA) and reveal the basic science of human life. This knowledge will make it possible to ameliorate, cure or prevent genetic diseases, enhancing the quality of life and decreasing healthcare costs to society. The fruits of this human biology research will undoubtedly carry over into agriculture, environmental management and industry.

Beyond its primary focuses, Livermore will continue to support other innovative science and technology initiatives that have the potential for high impact in their field and that reinforce scientific and technological strengths. In addition, Livermore remains committed to fostering science and math education to help ensure the scientific literacy of the general population and to inspire future generations of scientists and engineers.

To realize this vision, Livermore will rely on its excellent and diverse staff. Robust quality management systems will be put in place to ensure that laboratory operations are accountable, cost-effective, and meet measurable performance standards.

Livermore's vision for the future aligns with the business areas identified in the Department of Energy's new strategic plan: national security, energy resources, environmental quality, and industrial competitiveness — all addressed through science and technology. Livermore will build on and enhance partnerships with DOE staff to ensure excellence in the achievement of common goals.

Attaining Livermore's goals will also require new forms of cooperation among the national laboratories, universities, and industry. Key national facilities will be built and used by multi-institutional teams of researchers. Laboratory sites will be readily accessible to outside partners. The commercialization of new technologies will be the planned end product of these collaborative projects.

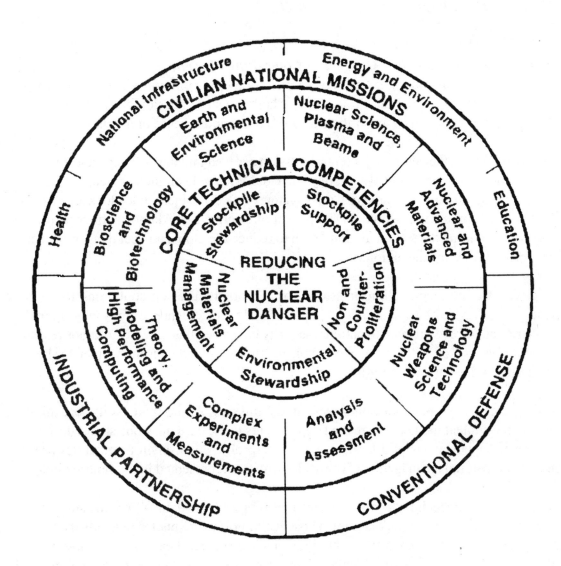

Figure 9-2. *The Los Alamos Mission. Courtesy of the U.S. Department of Energy.*

The vision for the future of the Livermore Laboratory is one of sustained, results-oriented excellence. Livermore is committed to serving the country as a national resource of scientific and technological expertise, dedicated to global security, the environment, and the future scientific needs of the nation.

Los Alamos National Laboratory (LANL)
(http://www.lanl.gov/)

Reducing the Global Nuclear Danger: A Compelling Mission
Los Alamos National Laboratory was born with a compelling mission. Created during World War II as part of the Manhattan Project, Los Alamos' central mission originally was to develop the first atomic bombs under the wartime's tremendous sense of urgency.

After World War II, the laboratory's central mission evolved into developing nuclear weapons for deterrence during the Cold War. Today, the Cold War cycle of nuclear weapons development has ended, and so the new central mission must reflect the incredible global events of the recent past. In conjunction with the Department of Energy, Los Alamos defines its primary mission as providing the technical foundation to reduce the global nuclear danger to ensure a more secure future for the nation.

Along with having a compelling mission, Los Alamos focuses on continuing to do great science in the service of the nation. By continuing to do great science — and many significant discoveries not directly related to national security have been made by Los Alamos researchers over the years — it will continue to be able to attract talented men and women to shore up its scientific and technical capabilities.

The diagram in *Figure 9-2* sums up Los Alamos' mission. At the center is its core mission: reducing the nuclear danger. There are five elements to the core mission: stockpile support, stockpile management, nonproliferation and counterproliferation, environmental stewardship, and nuclear materials management. These elements are represented in the central circle.

There are eight core technical competencies that support Los Alamos' core mission and surround the central circle on the diagram. Those technologies are nuclear and advanced materials; nuclear weapons science and technology; nuclear science, beams and plasmas; analysis and assessment; complex experimentation and measurements; theory, modeling and high-performance computing; bioscience and biotechnology; and earth and environmental sciences.

These areas of technical competency, in turn, provide the basis for Los Alamos to participate in civilian national missions (such as the Human Genome Project), conventional defense research and development, and industrial partnerships — represented in the outer ring of the diagram. Work in these areas serves important national needs while also supporting the laboratory's core mission by strengthening and maintaining core technical competencies.

Profile and History

Los Alamos National Laboratory was established in 1943 as Project Y of the Manhattan Engineering District to develop the world's first atomic bomb, under the leadership of J. Robert Oppenheimer. Today, Los Alamos is a multidisciplinary, multiprogram laboratory whose central mission still revolves around national security.

Managed since its beginning by the University of California, where Oppenheimer was a professor, Los Alamos continues a commitment to maintaining a tradition of free inquiry and debate, which is essential to any scientific undertaking. Located on the Pajarito Plateau about 35 miles northwest of Santa Fe, the capital of New Mexico, Los Alamos is one of twenty-eight Department of Energy laboratories across the country.

The laboratory's original mission to design, develop, and test nuclear weapons has broadened and evolved as technologies, U.S. priorities, and the world community have changed. Today, Los Alamos uses the core technical competencies developed for defense and civilian programs to carry out both national security responsibilities and its broadly based programs in energy, nuclear safeguards, biomedical science, environmental protection and cleanup, computational science, materials science, and other basic sciences. The capabilities resident in these programs are increasingly being used in partnership with industrial firms to bring laboratory-developed technology to the assistance of the overall competitiveness of the U.S. economy.

The laboratory fills an intermediate role — between academic research and industrial production — that helps expedite the development and commercialization of emerging technologies. In all of its programs, Los Alamos continues to maintain an intellectual environment that is open to new ideas. In addition, Los Alamos is committed to ensuring that all of its activities are designed to protect employees, the public, and the environment.

To accomplish these varied and challenging programs, the laboratory relies on technical staff from across the United States. Approximately one-third of the technical staff members are physicists, one-fourth are engineers, one-sixth are chemists and materials scientists, and the remainder work in mathematics and computational science, biological science, geoscience, and other disciplines. Professional scientists and students also come to Los Alamos as visitors to participate in scientific projects. The staff collaborates with universities and industry in both basic and applied research to develop resources for the future.

Los Alamos National Laboratory covers more than 43 square miles of mesas and canyons in northern New Mexico. As the largest institution and the largest employer in the area, the laboratory has approximately 7,000 University of California employees plus approximately 3,500 contractor personnel. Its annual budget is approximately $1 billion.

National Renewable Energy Laboratory (NREL)
(http://info.nrel.gov/)

NREL is the nation's leading laboratory for renewable energy and energy efficiency research. It develops renewable energy technologies to heat and cool buildings; increase efficiency in industry; light homes and offices; power cars and trucks; produce plastics, clothing and chemicals; clean water; and destroy toxic wastes. NREL's state-of-the-art research facilities are used to develop technologies for converting sunlight into electricity, advancing photovoltaic technology, making fuels and chemicals, developing cleaner fuels for vehicles, harnessing the wind, and for hydrogen research.

NREL's Solar Energy Research Facility
14975 Denver West Parkway, Golden, CO 80401
Completed in 1993, the Solar Energy Research Facility (SERF) houses 42 laboratories for research in photovoltaics, superconductivity and related materials sciences. SERF is one of the federal government's most energy efficient buildings, using approximately 30% less energy than typical research facilities.

NREL's Field Test Laboratory Building
15003 Denver West Parkway, Golden, CO 80401
Constructed in 1985, the Field Test Laboratory Building houses 41 laboratories for research in alternative fuels and industrial processes.

NREL's Denver West Office Park
1617 Cole Boulevard, Golden, CO 80401
NREL leases 275,000 square feet of office space in the Denver West Office Park. The space is used for administrative offices and research labs.

NREL's National Wind Technology Center
18200 State Highway 128, Golden, CO 80403
The National Wind Technology Center, located in the Rocky Flats Environmental Technology Site buffer zone, is the site of NREL's wind energy research.

NREL's Alternative Fuels User Facility

15137 Denver West Parkway, Golden, CO 80401

The recently completed Alternative Fuels User Facility provides laboratory space for methanol, ethanol and other alternative fuel research. It includes a pilot plant for evaluating biofuels production technologies.

NREL's Visitors Center

14869 Denver West Parkway, Golden, CO 80401

NREL's new Visitors Center features interactive exhibits and displays explaining renewable energy and energy efficiency technologies. It is open weekdays from 8 a.m. to 5 p.m., and during the summer, on weekends from 10 a.m. to 4 p.m.

NREL's Washington, D.C. Office

409 12th Street, SW., Suite 710, Washington, DC 20024-2125

NREL's Washington Office provides analysis, program and communications support for the laboratory in the metropolitan Washington area.

NREL's Outdoor Test Facility

15883 Denver West Parkway, Golden, CO 80401

The Outdoor Test Facility provides space for performing outdoor photovoltaics experiments.

NREL's Planned Facilities

The Thermal Test Facility, scheduled for completion in 1996, will provide research space for thermal testing of ventilation, and active and passive solar systems. NREL will request funding from DOE in 1997 for the Renewable Fuels and Chemical Facility and the Industrial Thermochemical User Test Facility.

Oak Ridge National Laboratory (ORNL)

(http://www.ornl.gov/)

ORNL is a Department of Energy multiprogram laboratory managed by Lockheed Martin Energy Research Corporation. Scientists and engineers at ORNL conduct a wide range of basic and applied research and development to advance (in several Core Competencies) the nation's energy resources, environmental quality, scientific knowledge, educational foundations, and economic competitiveness. ORNL collaborates with other federal agencies, industries, and universities. It is committed to excellence in all of its activities and operates in compliance with environmental, safety, and health laws and regulations.

Core Competencies
1. Energy production and end-use technologies.
2. Biological and environmental science and technology.
3. Advanced materials synthesis, processing, and characterization.
4. Neutron-based science and technology.
5. Computational science and advanced computing.

Overview of Oak Ridge National Laboratory Technologies
The mission of Oak Ridge National Laboratory is to advance the frontiers of science and technology in three broad arenas: energy, the environment, and economic competitiveness. In pursuit of this mission, ORNL conducts basic and applied research in these key areas: materials sciences and engineering; physical, chemical, and engineering sciences; biological and life sciences; computational sciences, and manufacturing sciences and technologies.

Energy
The world uses ten times more energy today than only a century ago. Industrialization, coupled with the world's current population growth—90 million more people every year—will strain global energy resources seriously in coming decades.

The 1991 war in the Persian Gulf underscores the risks of oil dependence. Coal, while plentiful, is difficult to mine safely and burn cleanly. Nuclear fission, hailed 30 years ago as the ideal power source, is beset by safety concerns, spiraling costs, and the thorny problem of waste disposal. Fusion, a potentially limitless source of energy, remains elusive. Conservation and efficiency requires technology that is too complex or expensive for much of the world. Renewable energy sources are intermittent (solar and wind power), limited (biomass and hydro), or expensive.

Conservation and Renewables
A few years ago, conservation was dismissed as "shivering in the dark." No more. Now equated with efficiency, it's the simplest way to increase industrial competitiveness, reduce dependence on imported oil, reduce pollution, lower energy costs, and extend the world's finite resources.

At the High Temperature Materials Laboratory, researchers from throughout the nation are helping to develop tough new ceramics for advanced diesel and gas-turbine engines. By combining hotter operation with light weight and great durability, a ceramic gas-turbine engine could boost the efficiency of today's engines dramatically.

Oak Ridge also plays a key role in saving energy used to heat and cool buildings. ORNL's Roof Research Center, jointly sponsored by the roofing and building-materials industries, is the nation's most sophisticated laboratory for testing energy-efficient materials and designs. The facility's large-scale climate simulator allows years of weathering to be compressed into weeks; it also allows precise measurements of how temperature, moisture, and other factors affect the energy performance and lifetime of roofs. Oak Ridge is also developing sophisticated heating and cooling systems that outperform today's models—without relying on ozone-destroying refrigerants.

Another promising route to energy efficiency is superconductivity, which eliminates losses from electric resistance in motors, generators, and power lines. Working with electric utilities and other industries, the Department of Energy and ORNL have established a Superconductivity Pilot Center here to help transform superconductivity into a widespread commercial technology.

At another end of the technological spectrum, ORNL is exploring new ways to produce and use that most basic of fuels: wood. Through advances in plant genetics, cultivation, and processing, ORNL is pioneering high-yield "energy crops" to fuel power plants and vehicles.

Fission
Without fundamental changes, commercial nuclear power may not survive beyond today's reactors. Spiraling costs, construction and licensing delays, and deep fears about accidents and waste have brought U.S. nuclear plant construction to a halt. Ironically, most energy analysts still agree that more fission power will be vital in meeting the nation's future energy needs.

As today's reactors age, ORNL works with the U.S. Nuclear Regulatory Commission to ensure that safety and reliability remain high. ORNL's Nuclear Safety Information Center is the national repository of data from existing nuclear plants. This data base logs the details of thousands of incidents each year; its analyses can help flag safety problems before they become serious. ORNL also conducts extensive research to verify the strength of reactor pressure vessels and other components under the most severe conditions.

Looking to the future, Oak Ridge engineers and researchers are helping to develop a new and safer generation of advanced reactors: modular, high-temperature gas-cooled reactors. Unlike today's reactors, they will be standardized, not custom-designed — an important advantage in cost, reliability, and the speed of construction and licensing. Most important, the reactors are designed to be passively safe. Their safety is designed *in*, not added *on*; even if they were to lose all coolant, these reactors would not melt down, but would shut off and cool gradually without operator intervention.

Fusion

Harnessing nuclear fusion for peaceful power production is one of the greatest technical challenges of all time. First, there's the matter of heating hydrogen ions to more than a hundred million degrees centigrade. Then there's the matter of confining them and overcoming their reluctance to fuse. Finally comes the really tricky part: extracting heat (and adding new fuel) without snuffing the fire.

ORNL's contributions to fusion include work in fusion physics using the world's largest "stellarator"; tests of giant superconducting magnets for confining the gas plasma where fusion occurs; plasma-heating systems which produce the starlike temperatures needed; "pellet injectors" which shoot droplets of frozen hydrogen fuel into the plasma; and materials designed to survive the extreme conditions in fusion reactors.

Fossil

The chemistry that created petroleum occurred millions of years ago — but the chemistry that determines whether an oil well produces is rooted firmly in the present. If the pores of oil-producing rock are clogged, for example, the oil may remain trapped forever. Working with petroleum industry scientists, Oak Ridge chemists are probing the geochemical differences between a gusher and a dry well to help refine predictive models and improve tomorrow's oil-recovery systems.

A lump of coal may look like a simple chunk of matter, but it's actually extraordinarily complex, containing some of nature's most Byzantine macromolecules. Intertwined with the coal's hydrocarbon fuel are impurities and corrosives that can foul up the chemistry, equipment and cells. To get a clearer picture, chemists are developing new analyses based on *nuclear magnetic resonance*, the technique that also can delineate muscles, nerves, and organs in the body. In addition, ORNL researchers are improving the materials and processes needed to convert this abundant resource into clean, economical sources of liquid fuels, gases, and electricity.

ORNL: Planning for Tomorrow

As we approach a new century — and a new millennium — it is becoming increasingly clear that we are at an important crossroads. If we choose wisely among the paths radiating before us, the potential for global progress is greater than at any other time in history. If we choose poorly, the risk is equally great.

As a pioneering national laboratory, ORNL is already working to meet the science and technology needs of the future — needs stretching to the frontiers of knowledge and understanding, but rooted in the lives of ordinary people. Clean air and water. Warmth and light. The hope for full, healthy lives.

Many of the choices and issues that will determine our future are interwoven with energy and the environment: oil supplies; greenhouse gases; a burgeoning world population; nuclear and hazardous waste; energy costs; industrial productivity and competitiveness; deforestation; transportation; the ozone layer; gene-damaging materials and mechanisms.

These, then, are the problems. These, too, are the opportunities — the urgent needs and challenges that can inspire scientists and engineers to new levels of hard work, creative thought, and technological can-do.

Between now and the year 2000, ORNL's research agenda will be shaped by several pressing issues:

1. *Energy needs*. After years of national neglect, energy is now reemerging as a critical priority. Developing nations, increasingly the centers of high population growth and low-cost manufacturing, will dominate future demand for oil, gas, and other resources. American nuclear power plants are aging, but no new plants have been licensed to replace them. Also, war in the Mideast has added a new urgency to the need for energy security, efficiency, and conservation. Scientists, engineers, and policy experts at ORNL are working to shape energy options and policies that are sensible, comprehensive, and visionary.

2. *Global environmental change*. Ozone depletion, deforestation, acid rain, the earth-warming "greenhouse effect," and waste disposal must be addressed as the stresses on our globe increase. How do we balance industrial development with environmental protection? How can we use energy and dispose of wastes with minimal harm? What strategies and policies offer the best hope for stopping the destruction of the earth's protective ozone layer? To explore questions such as these, ORNL has established a new Center for Global Environmental Studies. There, scientists from many fields are beginning to tackle the array of problems threatening the delicately-balanced biosphere we call home.

3. *The mysteries of life and matter.* Locked away in the intricate coding of DNA are the miracles of human development and the tragedy of hereditary disease. Using sophisticated new research techniques, ORNL geneticists are helping unlock these secrets and other fundamental mysteries of life. In the physical realm, chemists, physicists, and materials scientists are probing deeper into the essential nature of matter and learning to engineer entirely new materials for supercomputers, superconductors, high-efficiency engines, and other innovations. By the turn of the century, ORNL's Advanced Neutron Source, a powerful research reactor now in the design stages, could offer scientists from throughout the world an unequaled array of tools for exploring the structure of matter.

4. *Technological competitiveness.* ORNL is committed to helping restore the U.S. lead in science and technology. In Oak Ridge, progress gives rise to more progress; innovation breeds innovation. Steady advances in fields such as genetics, materials, computing, robotics, and intelligent machines are laying the foundations for future advances, including perhaps some that are beyond our present imagining.

5. *Education.* In recent years, DOE has elevated education to a top priority for the national laboratories. In doing so, it designated ORNL and several other labs as Science Education Centers. Today, ORNL's learn-by-doing educational programs reach out to more than 20,000 people each year. Programs tailored for students and teachers of all ages, from kindergarten through postgraduate research, represent an enormous investment in the scientists and engineers of the next century.

In addition to these priorities, ORNL is also stressing a new "style" of research and development — one that prizes collaboration, cooperation, and the sharing of technology and knowledge. This new approach will ensure that the nation's investment in research and development pays off in the U.S. and world markets... and in the lives of people for generations to come.

Pacific Northwest National Laboratory (PNL)
(http://www.pnl.gov/)

The Pacific Northwest National Laboratory, located at Richland and Sequim, WA, is one of nine U.S. Department of Energy multiprogram national laboratories. The laboratory employs 3,700 and has an annual budget of more than $500 million. Pacific Northwest National Laboratory conducts research in nearly every field of basic science to solve problems in the areas of the environment, energy, health and national security. Currently, laboratory staff are engaged in 2,000 research projects for the DOE, other federal agencies, and private clients.

The Pacific Northwest Laboratory's core mission is to deliver environmental science and technology in the service of the nation and humanity. Through basic research PNL creates fundamental knowledge of natural, engineered, and social systems that is the basis for both effective environmental technology and sound public policy. PNL solves legacy environmental problems by delivering technologies that remediate existing environmental hazards, addressing today's environmental needs with technologies that prevent pollution and minimize waste, and is laying the technical foundation for tomorrow's inherently clean energy and industrial processes. PNL also applies its capabilities to meet selected national security, energy, and human health needs, to strengthen the U.S. economy, and to support the education of future scientists and engineers.

Battelle Memorial Institute operates PNL as a multiprogram laboratory for the Department of Energy (DOE). It is committed to the principles of total quality to sustain excellence in research and development and the conduct of all laboratory operations, such as compliance with all applicable environmental, safety, and health requirements.

PNL acts as a steward of the DOE's resources by anticipating future national needs and investing in the development of major new capabilities and programs to meet these needs. Current investments include molecular science research, advanced processing technology, biotechnology, global environmental change research, and energy technology development.

PNL History
In 1965, Battelle Memorial Institute assumed management and operation of the federal government's Hanford Laboratories in southeastern Washington state. At the same time, the research facility was separated from Hanford site operations and renamed the Pacific Northwest Laboratory. Battelle invested $50 million in private research facilities adjacent to the government laboratory.

Initially, research at PNL focused on nuclear technology and the environmental and health effects of radiation. Today, the PNL staff conducts a spectrum of projects and programs in energy, the environment, and the economy. For example, PNL helped the Department of Energy establish the first Atmospheric Radiation Measurement site to obtain data related to global environmental change. PNL developed and transferred a thermochemical technology to industry that converts organic wastes to methane and carbon dioxide. PNL discovered ancient microbial organisms that hold potential for remediating contaminated soils deep within the earth.

One of the provisions of Battelle's contract with DOE is a unique agreement called a *Use Permit*. This agreement combines Battelle-owned and government-owned facilities in a consolidated laboratory where Battelle can conduct work for the DOE as well as other government agencies and private businesses on a cost-reimbursable basis. Each year, between 300 and 400 development contracts are conducted at PNL for industrial partners. In 1993, these contracts were valued at more than $14 million.

The *Use Permit* also gave a strong jumpstart to PNL's technology transfer program. Since 1965, the laboratory has routinely transferred technology — obtaining patents, licensing technology, and developing partnerships with industry. Today, PNL has more than 530 active licenses and other intellectual property agreements, and royalties from intellectual property reached $1 million in 1993. Three of the earliest patents issued to PNL staff contained the critical design elements for compact discs and CD players manufactured and marketed worldwide.

Sandia National Laboratories (SNL)
(http://www.sandia.gov/)

Research and Development Programs at Sandia
Sandia realizes its strategic intent of exceptional service in the national interest by supporting the Department of Energy's mission objectives: national defense, energy security, environmental integrity, and economic competition through a strong base in science and technology.

National Security Programs (Including Defense Programs Sector)
These programs for DOE and DOD entail stewardship and development of an evolving nuclear weapons stockpile; implementation of new structures, and streamlined approaches to future nuclear weapon production requirements; assumption of some non-nuclear production responsibilities of certain DOE production facilities designated for closure; development of arms control, nonproliferation, and counterproliferation technologies, and analysis of intelligence; and restoration of contaminated Sandia sites in order to comply with applicable environmental laws and regulations.

Energy and Environmental Programs
These programs for the DOE and the Nuclear Regulatory Commission involve identification, development, and deployment of full systems solutions for safe, clean, and affordable energy options; and development of affordable and deployable environmental technologies to remediate DOE sites and for broader applications of environmentally conscious operations in industry.

Work Other Than for DOE

This work for DOD, other federal agencies, and nonfederal agencies, applies Sandia's capabilities to enhance the success of other federal government agencies and nonfederal entities (such as state and local governments, private industry, and universities) by providing trusted, objective, and independent technical counsel; and accelerates the application of integrated technology solutions to problems of national importance through partnerships with industry.

Sandia's Scientific and Technical Capabilities

Sandia's ability to build solutions to complex problems relies on the labs' fundamental scientific and technical strengths. Sandia's core competencies are the cornerstone of the labs' reputations and underlie Sandia's ability to provide exceptional service to many customers.

Sandia's multidisciplinary research foundations provide the scientific knowledge base upon which the labs depend:

1. Computational and Information Sciences.
2. Engineered Materials and Processes.
3. Engineering Science.
4. Microelectronics and Photonics.

New, advancing technologies are strategically important for Sandia's future:

1. Advanced Information Technology.
2. Distributed Information Technologies.
3. Advanced Manufacturing Technology.
4. Electronics Technology.
5. Electronic Components.
6. Pulsed Power Technology.

Built upon these foundations, Sandia provides a wide range of research, development, engineering and design capabilities in the fields of:

1. Aerospace and Aerodynamics.
2. Combustion Research.
3. Command, Control, and Surety.
4. Computing and Information Technology.
5. Computational Engineering.

6. Distributed Information Technologies.
7. Geographic Information Systems.
8. Visual Communications.
9. Development Testing.
10. Energetic Materials (Explosives) and Components.
11. Energy Technologies (Non-Nuclear).
12. Renewable Energy Technologies.
13. Geoscience and Geotechnology.
14. Geosciences and Geotechnology Center.
15. Intelligent Systems and Robotics.
16. Archimedes — Robotic Assembly Planning.
17. Robotics Cross Cutting and Advanced Technology.
18. Robotic Vehicle Range.
19. Manufacturing Technologies.
20. Manufacturing Technologies Center.
21. Integrated Manufacturing Technology Laboratory.
22. National Machine Tool Partnership.
23. Nuclear Energy Technologies.
24. Analysis Capabilities.
25. Experimental and Test Facilities.
26. Physics and Chemistry.
27. Chemical Vapor Deposition (CVD) Sciences.
28. Combustion Chemistry Laboratory.
29. Materials Sciences.
30. Radiation Effects.
31. Sensors and Monitoring Systems.
32. Cooperative Monitoring Center.
33. Synthetic Aperture Radar.
34. Strategic and Systems Analysis and Engineering.
35. Engineering of Complex Systems.
36. Process Systems Analysis and Engineering.
37. Risk Assessment/Management.
38. Transportation Technology.
39. Transportation of Hazardous Materials.
40. Transportation Systems.
41. Waste Management Technologies.

9-8. NATIONAL LABORATORY WORLD-WIDE WEB SITES

Advanced Computing Laboratory (ACL)
　　http://www.acl.lanl.gov

Ames Laboratory
　　http://www.ameslab.gov

Argonne National Laboratory (ANL)
　　http://www.anl.gov

Bonneville Power Administration
　　http://www.bpa.gov

Brookhaven National Laboratory
　　http://www.bnl.gov/bnl.html

Center for Computational Sciences
　　http://www.ccs.ornl.gov

Continuous Electron Beam Accelerator Facility (CEBAF)
　　http://www.cebaf.gov/cebaf.html

Environmental Measurements Laboratory (EML)
　　http://www.eml.doe.gov

Fermi National Accelerator Laboratory (Fermilab)
　　http://fnnews.fnal.gov

Grand Junction Projects Office (GJPO)
　　http://www.doegjpo.com

Idaho National Engineering Laboratory (INEL)
　　http://www.inel.gov

Kansas City Plant (Allied Signal, Inc.)
　　http://www.as.kcp.com

Ernest Orlando Lawrence Berkeley Laboratory (LBL)
    http://www.lbl.gov

Lawrence Livermore National Laboratory
    http://www.llnl.gov

LLNL High Energy Physics
    http://gem1.llnl.gov/home.html

Los Alamos National Laboratory (LANL)
    http://www.lanl.gov

Morgantown Energy Technology Center (METC)
    http://www.metc.doe.gov

National Energy Research Supercomputer Center
    http://www.nersc.gov

National Renewable Energy Lab (NREL)
    http://info.nrel.gov

Oak Ridge Associated Universities
    http://www.orau.gov

Oak Ridge National Laboratory
    http://www.ornl.gov

Pacific Northwest National Lab
    http://www.pnl.gov

Princeton Plasma Physics Lab
    http://www.pppl.gov

Sandia National Laboratories
    http://www.sandia.gov

Savannah River Site (Westinghouse)
    http://www.srs.gov

Stanford Linear Accelerator Center
   http://www.slac.stanford.edu

Superconducting Super Collider (SSC)
   http://www.ssc.gov

University of California (Lab Management Home Page)
   http://labs.ucop.edu

EPILOGUE

Now that you have an overall understanding of alternatives and renewables, and the infrastructure that now exists, what will you do with the knowledge? Let me suggest that you continue to educate yourself in these areas. Use the addresses that I have provided to you (in the appendices) to contact the U.S. Department of Energy, various industry associations, and companies to get on mailing lists and gather more information. If you are connected to the Internet, visit the many sites that I have listed throughout the book. My site directory, accessible from my home page, lists them all, and I try to keep them updated. It is an understatement to say you will find a wealth of information on the World-Wide Web. If you are a teacher, include these subjects in your curriculum and teach your students what you have learned here. More importantly, teach them how to research the topics to gather more information.

In addition to keeping yourself informed, make your voice known by contacting your government representatives at all levels. Let them know that we must take renewables seriously and that you expect them to favor tax incentives and cooperative grants for research and corporate expansion. Let them know that we do not want a future of dependence on OPEC nations for oil at high cost. A world dependent on OPEC oil will be a world filled with tension and distress. Remember, too, that even OPEC oil sources will diminish in the 21st century.

Finally, see how you can apply renewable energy sources to your life. Consider using photovoltaics or wind to generate some or all of your electricity for your home. Use the sun to heat water. Learn to conserve energy whenever possible. Invite your local utility company to analyze your home for energy efficiency; they will be able to make several energy-saving suggestions. Begin to adjust now to a life style different than the excesses we have enjoyed. Starting to prepare now will ease the shock of the ever-increasing gasoline and electricity costs that will affect us all as we move into the 21st century.

Encourage automakers to make available (at reasonable cost) electric vehicles to meet our commuting needs. Since most oil production goes to vehicle use, the main focus must be placed on new energy sources for transportation. Many advances have been made in electric vehicle design and batteries used for energy storage; but, we must be ready to accept these new vehicles. Our responsibility will be to change our attitudes regarding excessive power under the hood. If we are not willing to buy the new electric vehicles, the manufacturers will not offer them. However, as time goes on, we will see that we have little choice with increasing gasoline and oil prices. Our decision will be a practical one. The catch here will be if the average person is able to afford to *buy* a new electric vehicle. The current price of an electric vehicle now being offered by a major automobile manufacturer leaves one wondering who can afford to buy it. However, there are other companies that have sprung up around the world who manufacture electric vehicles at more reasonable prices. We will hear more about them as gas prices rise. Again, if you are on the Internet, I have a listing regarding electric vehicles on my site directory, accessible from my home page. If you wish to contact me, you may do so from my home page: http:www.castlegate.net/personals/hazen

A

APPENDIX: ACRONYMS & SYMBOLS

Å:	Ampere
A:	Angstrom = 1×10^{-10} Meters
AC:	Alternating Current
ACL:	Advanced Computing Laboratory
ALCAN:	Aluminum of Canada
AMOCO:	American Oil Company
ANL:	Argonne National Laboratory
ARCO:	Atlantic Richfield Company
a-Si:	Amorphous Silicon
AWEA:	American Wind Energy Association
AWT:	Advanced Wind Turbines, Inc.
BEST:	Battery Energy Storage Test Facility
BR:	Breeder Reactor
BTU:	British Thermal Unit
BWR:	Boiling Water Reactor
CCPA:	Central California Power Agency
CdTe:	Cadmium Telluride
CEBAF:	Continuous Electron Beam Accelerator Facility
CECRI:	Central Electrochemical Research Institute
CuInSe:	Copper Indium Diselenide
CuInGaSe:	Copper Indium Gallium Diselenide
CVD:	Chemical Vapor Deposition
D:	Deuterium
DC:	Direct Current
DOE:	Department of Energy (U.S.)
E:	Electromotive Force or Energy
EHD:	Extended Height to Diameter Ratio
EML:	Environmental Measurements Laboratory
EPRI:	Electric Power Research Institute
ERDA:	Energy Research and Development Administration
FNAL:	Fermi National Accelerator Laboratory

HAWT:	Horizontal-Axis Wind Turbine
hp:	Horsepower
hph:	Horsepower-Hour
http:	Hypertext Transfer Protocol
Hz:	Hertz
I:	Intensity of Flow
INEL:	Idaho National Engineering Laboratory
J:	Joule
kHz:	Kilohertz = 1000 Cycles Per Second
km/h:	Kilometers Per Hour
kWh:	Kilowatt-Hour
LANL:	Los Alamos National Laboratory
LBL:	Lawrence Berkeley Laboratory
LLNL:	Lawrence Livermore National Laboratory
LMFBR:	Liquid Metal Fast Breeder Reactor
LWR:	Light Water Reactor
m:	Meters
MHz:	Megahertz = 1,000,000 Cycles Per Second
MPH:	Miles Per Hour
NASA:	National Aeronautics and Space Administration
NCPA:	Northern California Power Agency
NELHA:	Natural Energy Laboratory of Hawaii Authority
NRC:	Nuclear Regulatory Commission (U.S.)
NREL:	National Renewable Energy Laboratory
ORNL:	Oak Ridge National Laboratory
OTEC:	Ocean Thermal Energy Conversion
P:	Power
PG&E:	Pacific Gas & Electric
PICHTR:	Pacific International Center for High Technology Research
PNL:	Pacific Northwest National Laboratory
p-Si:	Polycrystalline Silicon
Pu:	Plutonium
PV:	Photovoltaic
PWR:	Pressurized Water Reactor
R:	Resistance
RPM:	Revolutions Per Minute
SA:	Stand Alone

Si:	Silicon
SMUD:	Sacramento Utility District
SNL:	Sandia National Laboratories
SSC:	Superconducting Super Collider
STC:	Solar Thermal Conversion
TEC:	Thermoelectric Cooling
TFT:	Thin-Film Technology
U:	Uranium
μ:	Micro = 0.000001
UI:	Utility-Interactive or Utility Integrated
UNOCAL:	Union Oil Company of California
μV:	Microvolt
UWIG:	Utility Wind Interest Group
V:	Volt
VAWT:	Vertical-Axis Wind Turbine
W:	Watt
Wp:	Peak Watts
WWW:	World-Wide Web
x-Si:	Single-Crystal Silicon

B

APPENDIX: ADDRESSES

ENERGY (MISC.)

Alternative Energy Engineering, Inc.
P.O. Box 339
1155 Redway Drive
Redway, CA 95560
(800) 777-6609
Fax: (800) 777-6648
http://www.alt-energy.com/aee/
http://www.asis.com/aee/

DO IT HOMESTEAD
Smith Mesa
P.O. Box 852
La Verkin, Utah 84745
(801) 877-1061
http://www.netins.net/showcase/solarcatalog/

Home Power Magazine
P.O. Box 520
Ashland, OR 97520
(916) 475-3179
Bulletin Board: (707) 822-8640
http://www.homepower.com

Hutton Power Systems Group
5470 Oakbrook Parkway, #G
Norcross, GA 30093
(800) 741-3811
Fax: (770) 729-9567
E-mail: pvphoton@aol.com

National Technical Information Service
U.S. Department of Commerce
5285 Port Royal Road
Springfield, VA 22161
(703) 487-4650

GEOTHERMAL

Geothermal Resources Council
P.O. Box 1350
Davis, CA 95617
(916) 758-2360

Naval Weapons Center
Geothermal Program Office
Naval Air Weapons Station
Code 823GØØD
1 Admin. Circle
China Lake, CA 93555-6100
(619) 939-2700

Pacific Gas and Electric Co.
Geysers Power Plant
P.O. Box 456
Healdsburg, CA 95448
(707) 431-6000
Fax: (707) 431-6066

NATIONAL LABORATORIES

Argonne National Laboratory
9700 South Cass Avenue
Argonne, IL 60439-4815
(800) 627-2596 and (708) 252-6489

National Renewable Energy Laboratory
1617 Cole Boulevard
Golden, CO 80401-3393
(303) 275-4090
Fax: (303) 275-4091
http://www.nrel.gov/

Oak Ridge National Laboratory
P.O. Box 2008
Oak Ridge, TN 37831-6486
(423) 576-1955
Fax: (423) 241-5722

Sandia National Laboratories
P.O. Box 5800
Albuquerque, NM 87185-0702
(505) 845-8221
Fax: (505) 844-0591
http://www.sandia.gov/renewable_energy/

U.S. Department of Energy
Office of Utility Technologies
1000 Independence Ave.
Washington, DC 20585
(202) 586-9220
http://www.doe.gov/
For information requests:
Energy Efficiency & Renewable Clearinghouse
Eric or Eren
P.O. Box 3048
Merrifield, VA 22116
(800) 363-3732
(703) 833-0400
http://www.eren.doe.gov/

OCEAN THERMAL ENERGY CONVERSION

Natural Energy Laboratory of Hawaii Authority
73-4460 Kaahumanu Hwy., #101
Kailua-Kona, HI 96740
(808) 329-7341
Fax: (808) 326-3262
http://bigisland.com/nelha

Sea Solar Power, Inc.
2422 S. Queen St.
York, PA 17402
(717) 741-0884
Fax: (717) 741-0886

SOLAR: PHOTOVOLTAIC

EPRI Electrical Power Research Institute
3412 Hillview Avenue
P.O. Box 10412
Palo Alto, CA 94303-0813USA
(415) 855-2000
http://www.epri.com

Independent Energy
The Industry's Business Magazine
620 Central Avenue
North Milaca, MN 56353-1788

National Renewable Energy Laboratory
1617 Cole Boulevard
Golden, CO 80401-3393

Sandia National Laboratories
Division 223
P.O. Box 5800
Albuquerque, NM 87185

Solar Energy Industries Association
122 "C," St. NW
4th Floor
Washington, DC 20001
(202) 383-2600

Solar Industry Journal
c/o Solar Energy Industries Association
122 "C," St. NW
4th Floor
Washington, DC 20001

Solar Today Magazine
American Solar Energy Society
2400 Central Avenue, Suite G-1
Boulder, CO 80301
(303) 443-3130
http://www.ases.org/solar

SOLAR: PHOTOVOLTAIC - DEALERS

Atlantic Solar Products
P.O. Box 70060
Baltimore, MD 21237
(410) 686-2500

Fowler Solar Electric
P.O. Box 435
Worthington, MA 01098
(413) 238-5974

Inter Island Solar
345 North Nimitz Hwy.
Honolulu, HI 96817
(808) 523-0711

Michigan Energy Works
9605 Potters Road
Saranac, MI 48881
(616) 897-5161

Real Goods
555 Leslie St.
Ukiah, CA 95482
(707) 468-9214

Remote Power
1608 Riverside
Fort Collins, CO 80524
(970) 482-9507

Solar Design Associates
P.O. Box 242
Harvard, MA 01451
(508) 456-6855
E-mail: sda@solardesign.com

Solaray Systems, Inc.
168 West Dearborn Street
Englewood, FL 34223
(941) 474-2421

SOLAR: PHOTOVOLTAIC - MODULES

Advanced Photovoltaic Systems
P.O. Box 7093
Princeton, NJ 08543-7093
(215) 343-7080
Fax: (215) 343-7081

AstroPower, Inc.
Solar Park
Newark, DE 19716-2000
(302) 366-0400
Fax: (302) 368-6474

Kyocera America
8611 Balboa Avenue
San Diego, CA 92123
(619) 576-2647
Fax: (619) 467-1487
http://www.kyocera.com/

Siemens Solar Industries
P.O. Box 6032
4650 Adohr Lane
Camarillo, CA 93011
(805) 388-6289
Fax: (805) 388-6395
http://www.solarpv.com/solarpv

Solarex Corporation
630 Solarex Court
Frederick, MD 21703
(800) 521-7652
Fax: (301) 698-4201

United Solar Systems Corp.
1100 West Maple Road
Troy, MI 48084
(810) 362-3120
(800) 843-3842
Fax: (810) 362-4442

SOLAR THERMAL

Sandia National Laboratories
Solar Thermal Technology Department
P.O. Box 5800
Albuquerque, NM 87185-0703
(505) 844-0195
Fax: (505) 844-7786

WIND

American Wind Energy Association
122 C Street, NW
4th Floor
Washington, DC 20001-2109
(202) 383-2500
Fax: (202) 383-2505
E-mail: windmail@mcimail.com

Advanced Wind Turbines, Inc.
425 Pontius Avenue North, Suite 150
Seattle, WA 98109-5450
(206) 292-0070
Fax: (206) 292-0075

FloWind Corporation
990 A Street, Suite 300
San Rafael, CA 94901
(415) 721-1300
Fax: (415) 721-1325

Bergey Windpower Company
2001 Priestley Avenue
Norman, OK 73069
(405) 364-4212
Fax: (405) 364-2078
E-mail: mbergey@bergey.com
http://www.bergey.com

Kenetech Windpower
Mr. William J. Whalen
500 Sansome Street, Suite 600
San Francisco, CA 94111-3211
(415) 398-3825
Fax: (415) 391-7740

Kern Wind Energy Association
P.O. Box 277
Tehachapi, CA 93581
(805) 822-7956
Fax: (805) 822-8452
E-mail: kweawhite@aol.com

Northern Power Systems
A New World Power Technology Company
1 North Wind Road
Moretown, VT 05660
(802) 496-2955
Fax: (802) 496-2953

Southwest Windpower, Inc.
2131 N. First Street
Flagstaff, AZ 86004
(520) 779-9463
Fax: (520) 779-1485
E-mail: swwindpower@ichange.com

WINDTech International, L.L.C.
1827 Powers Ferry Road
Building 2
Atlanta, GA 30339
(770) 933-2373
Fax: (770) 984-1869

Wind Turbine Industries Corporation
16801 Industrial Circle SE
Prior Lake, MN 55372
(612) 447-6049
Fax: (612) 447-6050

World Power Technologies, Inc.
19 N. Lake Avenue
Duluth, MN 55802
(218) 722-1492
Fax: (218) 722-0791
E-mail: wpt@cp.duluth.mn.us
Home Page: http://www.webpage.com/wpt

Zond Systems, Inc.
13000 Jameson Road
P.O. Box 1910
Tehachapi, CA 93581
(805) 822-6835
Fax: (805) 822-7880

C

APPENDIX: ANSWERS TO QUESTIONS

CHAPTER 1

1. Electromagnetic, electrochemical, thermoelectric, piezoelectric, photoelectric.
2. A lead-acid car battery.
3. Magnetic field, conductor in the magnetic field, relative motion between the two.
4. A series-parallel combination of thermocouples.
5. Yes.
6. Pressure or sudden impact applied to a crystal produces a voltage pulse and voltage pulses applied to a crystal produces flexing of the crystal.
7. A sonic transducer produces or responds to audible sound pressure while the ultrasonic transducer operates above the range of human hearing.
8. Fetal monitoring, burglar alarms, gallstone treatment, etc.

CHAPTER 2

1. Electrical energy comes from other forms of energy through the use of a transducer.
2. Energy can neither be created nor destroyed.
3. *Energia* - Greek, meaning "activity." It is the capacity for doing work.
4. Kinetic energy is the energy contained in a moving mass.
5. 0.45 BTU x 0.000293 kWh/BTU = 0.0001318 kWh
   0.45 BTU x 1055 J/BTU = 474.8 J
6. $I = 24V/50\Omega = 0.48A$
7. $P = I^2 \times R = 4A^2 \times 10\Omega = 160W$
8. kWh = 100W x 4.5 hours = 450Wh = 0.45 kWh
9. Cost = Usage x Rate = 2 kW x 16 hours/day x 20 days x $0.12/kWh = $76.80
10. 340Vp x 0.707 = 240V = 240VAC = 240V$_{effective}$

CHAPTER 3

1. As electromagnetic waves.
2. Hydro energy as in hydroelectric, solar thermal energy, chemical energy stored in fossil fuels.

3. Roughly 950 Wp/m² and 1220 Wp/m² —1000 Wp/m² is often used in calculations.
4. Batteries —chemical energy storage.
5. 40%.
6. A large number of solar reflectors are used to concentrate the sun's energy on one object to create a large amount of heat.
7. Jet engines, diesel engines, Stirling engines, and steam engines.
8. Ocean Thermal Energy Conversion = OTEC.

CHAPTER 4

1. HAWT = horizontal-axis wind turbine such as a conventional windmill.
   VAWT = vertical-axis wind turbine such as an anemometer.
2. A synchronous inverter is a DC to AC converter that produces an AC that is synchronized with the AC line frequency so the line can be used for energy storage.
3. VAWT.
4. The batteries are very expensive and must be routinely cared for and eventually replaced (perhaps every 3 to 5 years). Another disadvantage is that of limited energy storage.
5. Wind availability is perhaps the most obvious and significant problem with wind systems. The solution to this problem is to place wind turbines in regions that have a history of sustained energy winds and to use wind energy in conjunction with other sources of energy. Wind constancy is another serious problem facing wind energy utilization. The solution to the problem of wind constancy is manifold and costly. Mechanical mechanisms called constant speed drives, or converters, are employed to help regulate alternator rotation speed under varying turbine speed conditions. Also, instead of a single extremely high power wind turbine, a network of relatively low power wind turbines can be erected over a large open plain. All units can be interconnected and computer controlled in order to help overcome wind variations.
6. The American Wind Energy Association (AWEA) is a trade and advocacy organization based in Washington, D.C. AWEA represents the U.S. wind energy industry and individuals who support clean energy in the legislative/regulatory, public relations, and international arenas.

CHAPTER 5

1. The ocean offers up large masses of water that must return from whence it came. Man intervenes with some clever device that causes the water to work on its return. Thus a wave is captured and the mass of the returning water, acted upon by gravity, is channeled and forced to turn a turbine shafted to a low speed generator or alternator.

2. One possible solution is the conversion of ocean energy to chemical energy by the production of hydrogen through electrolysis.

CHAPTER 6

1. Hot springs and geysers, and as dry steam
2. Hot rock geo-energy is closely related to hydrothermal energy in that extremely hot rock, deep in the earth's crust, is used to super heat water turning it into highly pressurized steam. In this case, however, the rock is naturally dry and water is pumped down from the surface to be super-heated.
3. AIR POLLUTION. Many hazardous gases such as hydrogen sulfide, carbon dioxide, methane, and ammonia can be released in the process. Methods have been and are being developed to capture these secondary products and put them to useful applications.
   WATER POLLUTION. Subterranean water contains many minerals including large quantities of salt. If this water is permitted to mix with surface waters, plants and animals will be destroyed and the local ecological system will be disrupted. The solution to this problem seems to be purification or reinjection of subterranean waters back into the underground reservoirs.

CHAPTER 7

1. Boiling water reactors (BWRs) and the pressurized water reactors (PWRs).
2. The nuclear reaction, either directly or indirectly, superheats water converting it to steam to be used in powering the turbines.
3. The splitting of the atom's nucleus, and therefore the atom, is known as fission.
4. LWRs.
5. Breeder reactors use U-238 and plutonium (Pu-239) instead of U-235.
6. Supercritical.
7. Pu-239.
8. Fusion is another type of nuclear reaction in which the nuclei of two atoms are forced together to form an atom of a different element. The nuclei are said to be fused together forming one larger nucleus.
9. Advantages of fusion: fuel source not rare or costly, no radiation or waste products
   Disadvantages of fusion: difficult to start reaction since very high temperature is required ($100,000,000°$ C), very high initial construction cost.

CHAPTER 8

1. Hydro-energy storage.
2. OTEC.
3. An alternative energy source can be used to power a high pressure pump to store a massive volume of compressed air. The compressed air can then be used to drive special air turbines which, in turn, drive electrical generators.
4. The flywheel can be charged and discharged an almost limitless number of times, unlike batteries. Maintenance is very minimal. There are no potentially dangerous byproducts, such as hydrogen released as lead-acid batteries are charged. The flywheel can be charged and discharged very quickly. Initial and operating costs are far less than other storage systems.

CHAPTER 9

No questions.

D

APPENDIX: BIBLIOGRAPHY

Anderson, Hilbert J., *The Sea Solar Power Plant at Kulasekarapattinam*, Sea Solar Power, Inc., December 12, 1995

Blickstein, Jay, "Cheaper Solar Cells with Polycrystalline Silicon," *Popular Science Magazine*, New York, NY, September 1985

Casler, David, "Solar Power for Your Ham Station," *QST Magazine*, April 1996

DiChristina, Mariette, "Sea Power," *Popular Science Magazine*, New York, NY, May 1995

DOE's Solar Thermal Electric Program, U.S. Department of Energy, Washington, D.C., April 1993

Energy, National Geographic Society, Washington, D.C., 1981

First Reactor, The 40th Anniversary, DOE/NE-0046, U.S. Department of Energy, Washington, D.C., December 1982

Fischetti, Mark A., "Inherently Safe Reactors: They'd work if we'd let them," *IEEE Spectrum Magazine*, New York, NY, April 1987

Fischetti, Mark A., "Turning Neutrons into Electricity," *IEEE Spectrum Magazine*, New York, NY, August 1984

Fischetti, Mark A., "When Reactors Reach Old Age," *IEEE Spectrum Magazine*, New York, NY, February 1986

Fisher, Arthur, "Cold Fusion", *Popular Science Magazine*, New York, NY, April 1987

Fowler, John M., *Electricity from the Sun - Solar Photovoltaic Energy*, Fact Sheet HCP/U 3841-11, U.S. Department of Energy, Washington, D.C., September 1978

Fowler, John and Kathryn, National Science Teachers Association, *Solar Sea Power - Ocean Thermal Energy Conversion*, Fact Sheet EDM-1043-6, U. S. Department of Energy, Washington, D.C., 1977

Fowler, John and Kathryn, National Science Teachers Association, *Wind Power*, Fact Sheet EDM-1043-3, U.S. Department of Energy, Washington, D.C., 1977

Hazen, Mark E., *Experiencing Electricity and Electronics*, 2nd Edition, Saunders College Publishing, Philadelphia, PA, 1993

Hazen, Mark E., *Exploring Electronic Devices*, Saunders College Publishing, Philadelphia, PA, 1991

In Review, National Renewable Energy Laboratory, Golden, CO, Spring 1994

In Review, National Renewable Energy Laboratory, Golden, CO, Summer 1994

In Review, National Renewable Energy Laboratory, Golden, CO, Fall 1994

In Review, National Renewable Energy Laboratory, Golden, CO, Spring 1995

In Review, National Renewable Energy Laboratory, Golden, CO, Summer 1995

In Review, National Renewable Energy Laboratory, Golden, CO, Fall 1995

Jayadev, Jay, "Harnessing the Wind," *IEEE Spectrum Magazine*, New York, NY, November 1995

Kastner, Jacob, *Nature's Invisible Rays*, U.S. Energy Research and Development Administration, Washington, D.C., 1973

Kocivar, Ben, "World's Biggest Windmill Turns on for Large-Scale Wind Power," *Popular Science Magazine*, New York, NY, March 1976

Lindsley, E. F., "Advanced Design Puts New Twist in Windmill Generator," *Popular Science Magazine*, New York, NY, October 1979

Lindsley, E. F., "33 Windmills You Can Buy Now!," *Popular Science Magazine*, New York, NY, July 1982

Making Electricity from Solar, Wind, and Water, Alternative Energy Engineering, Redway, CA, 1995-96

McDermott, Jeanne, "Whatever Happened to Solar Cells?," *Popular Science Magazine*, New York, NY, January 1984

News Release, April 19, 1996, *Wind Energy Weekly*, #681, American Wind Energy Association, April 1996

Nuclear Power from Fission Reactors, DOE/NE-0029, U. S. Department of Energy, Washington, D.C., March 1982

Ocean Thermal Energy Conversion, The Department of Business, Economic Development, & Tourism, State of Hawaii, 1995

Parker, Sybil, *Encyclopedia of Energy*, 2nd Edition, McGraw-Hill Book Company, New York, NY, 1981

Photovoltaics - the Power of Choice, National Renewable Energy Laboratory, U. S. Department of Energy, Washington, D.C., January 1996

Properties of Deep Sea Water, Natural Energy Laboratory of Hawaii Authority

Rapp, Donald, *Solar Energy*, Prentice-Hall, Inc., Englewood Cliffs, NJ, 1981

Rogers, James L., *Energy*, Sandia National Laboratories, Albuquerque, NM, April 1980

Schefter, Jim, "Barrel-Blade Windmill - Efficient Power from the Magnus Effect," *Popular Science Magazine*, New York, NY, August 1983

Shipments of Nuclear Fuel and Waste - Are They Really Safe?, DOE/EV-0004/2, UC-71, U.S. Department of Energy, Washington, D.C., August 1978

Solar Thermal Energy, U.S. Department of Energy, Washington, D.C., DOE/CS-0200, November 1980

Stover, Dawn, "The Forecast for Wind Power," *Popular Science Magazine*, New York, NY, July 1995

Stover, Dawn, "The Nuclear Legacy," *Popular Science Magazine*, New York, NY, August 1995

Strong, Steven J., *The Solar Electric House*, Sustainability Press, Still River, Mass., 1993

Sunscape, Siemens Solar Industries, Camarillo, CA, 1994

Truscello, Vincent C. and Davis, Herbert S., "Nuclear-Electric Power in Space," *IEEE Spectrum Magazine*, New York, NY, December 1984

U.S. Wind Energy Industry, The, American Wind Energy Association, Washington, D.C., February 1995

Zorpette, Glenn, "Photovoltaics: Technical Gains and an Uncertain Market," *IEEE Spectrum Magazine*, New York, NY, July 1989

E

APPENDIX: WEB SITES

AUTHOR'S HOME PAGE

http://www.castlegate.net/personals/hazen/

ELECTROCHEMISTRY

Central Electrochemical Research Institute (CECRI), India
http://chpc06.ch.unito.it/electrochemistry.html

The Electrochemistry Gateway
http://www.soton.ac.uk/~slt1/EchemGate.html

ELECTROMAGNETIC

GE Power Systems - Generators... An Overview
http://www.ge.com/geps/turbines/gen-hst.html

ENERGY

California Energy Commission
http://www.energy.ca.gov/energy/html/directory.html

Energy Story
http://www.energy.ca.gov/energy/education/story/story-html/story.html

Energy Yellow Pages
http://www.ccnet.com/~nep/yellow.htm

National Renewable Energy Laboratory (NREL)
http://www.nrel.gov/

ENERGY STORAGE

California Energy Commission
    http://www.energy.ca.gov/energy/html/directory.html

GEOTHERMAL ENERGY

CREST's Intro to Geothermal
    http://solstice.crest.org/renewables/geothermal/grc/index.html

Alan Glennon's Geyser Page
    http://www.wku.edu:80/~glennja/pages/geyser.html

BRIDGE — British Mid Ocean Ridge Initiative
    http://www.nwo.ac.uk/iosdl/Rennell/Bridge/

Coso Geothermal Project
250 megawatt Geothermal power plant in California
    http://www1.chinalake.navy.mil/Geothermal.html

Fourth International Meeting
Heat Flow and the Structure of the GeoSphere, June 10-16, 1996
    http://www.eps.mcgill.ca/~hugo/heat.html

Geothermal Education Office
    http://www.ensemble.com/geo

Geothermal Energy in Iceland
    http://www.os.is/os-eng/geo-div.html

Geothermal Exploration in Korea
    http://www.kigam.re.kr/env-geology.html

Geothermal Heat Pump Consortium
    http://www.ghpc.org/index.html

Geothermal Heat Pump Initiative in the U.S.
    http://www.eren.doe.gov/ee-cgi-bin/cc_heatpump.pl

Geothermal Resources Council (USA) Library & Information
   http://www.demon.co.uk:80/geosci/grclib.html

Geothermal Section
Course in Renewable Energy (Physics 162)
Part of the Electric Universe Project at University of Oregon
   http://zebu.uoregon.edu/ph162/l18.html

Geothermics
International Journal of Geothermal Research and Its Applications
   http://www.elsevier.nl/catalogue/SA2/230/03700/03770/389/389.Html

International Geothermal Association
   http://www.demon.co.uk:80/geosci/igahome.html

Rotorua, New Zealand, Geothermal Areas
   http://www.akiko.lm.com/NZ/NZTour/Rotorua/Geothermal.html

Stanford University Geothermal Program
   http://ekofisk.stanford.edu/geotherm.html

U.S. Department of Energy (DOE)
Energy Efficiency and Renewable Energy Network
   http://www.eren.doe.gov/ee_renen-geo.html

HYDROPOWER

Alternative Energy Engineering, Inc.
   http://www.alt-energy.com/aee

Alternative Energy Sources Course, University of Oregon
   http://zebu.uoregon.edu/1996/phys162.html

Centre for Alternative Technology
   http://www.foe.co.uk/CAT/index.html

Mr. Solar's Home Page
   http://www.netins.net/showcase/solarcatalog/

Solstice
   http://solstice.crest.org/

NATIONAL LABORATORIES

Advanced Computing Laboratory (ACL)
   http://www.acl.lanl.gov

Ames Laboratory
   http://www.ameslab.gov

Argonne National Laboratory (ANL)
   http://www.anl.gov

Bonneville Power Administration
   http://www.bpa.gov

Brookhaven National Laboratory
   http://www.bnl.gov/bnl.html

Center for Computational Sciences
   http://www.ccs.ornl.gov

Continuous Electron Beam Accelerator Facility (CEBAF)
   http://www.cebaf.gov/cebaf.html

Environmental Measurements Laboratory (EML)
   http://www.eml.doe.gov

Fermi National Accelerator Laboratory (Fermilab)
   http://fnnews.fnal.gov

Grand Junction Projects Office (GJPO)
   http://www.doegjpo.com

Idaho National Engineering Laboratory (INEL)
   http://www.inel.gov

Kansas City Plant (Allied Signal, Inc.)
http://www.as.kcp.com

Ernest Orlando Lawrence Berkeley Laboratory (LBL)
http://www.lbl.gov

Lawrence Livermore National Laboratory
http://www.llnl.gov

LLNL High Energy Physics
http://gem1.llnl.gov/home.html

Los Alamos National Laboratory (LANL)
http://www.lanl.gov

Morgantown Energy Technology Center (METC)
http://www.metc.doe.gov

National Energy Research Supercomputer Center
http://www.nersc.gov

National Renewable Energy Lab (NREL)
http://info.nrel.gov

Oak Ridge Associated Universities
http://www.orau.gov

Oak Ridge National Laboratory
http://www.ornl.gov

Pacific Northwest National Lab
http://www.pnl.gov:2080

Princeton Plasma Physics Lab
http://www.pppl.gov

Sandia National Laboratories
http://www.sandia.gov

Savannah River Site (Westinghouse)
http://www.srs.gov

Stanford Linear Accelerator Center
http://www.slac.stanford.edu

Superconducting Super Collider (SSC)
http://www.ssc.gov

University of California (Lab Management Home Page)
http://labs.ucop.edu

NUCLEAR ENERGY

California Energy Commission
http://www.energy.ca.gov/

Fusion Energy
http://FusionEd.gat.com/

Nuclear Energy
http://www.phoenix.net/~nuclear/univ.html

Nuclear Information WWW Server
http://nuke.westlab.com/

Nuclear Listings
http://www.energy.ca.gov/energy/earthtext/other.html#NUCLEAR

Oak Ridge National Laboratory
http://www.ornl.gov

Stanford Nuclear Energy FAQs
http://steam.stanford.edu/jmc/progress/nuclear-faq.html

Todd's Atomic Homepage
http://neutrino.nuc.berkeley.edu/neutronics/todd.html

U.S. Nuclear Regulatory Commission
   http://www.nrc.gov/

PHOTOELECTRIC

Sandia National Laboratories - Photovoltaic Technology
   http://www.sandia.gov/Renewable_Energy/photovoltaic/pv.html

NW Solar Cell
   http://www.seanet.com/Users/miknel/NW-SolarCell.html

Centre for Photovoltaic Devices and Systems
   http://www.vast.unsw.edu.au/pv.html

PIEZOELECTRIC

High-Temperature Pressure Transducers
   http://www.ssec.honeywell.com/papers/HTPT/HTPT.html

Teledyne Brown Engineering - Transducers and Sensors
   http://www.tbe.com/products/sensors/sensors.html

NASA - Instrument and Sensing Technology
   http://ranier.oact.hq.nasa.gov/sensors_page/insthp.html

PHOTOVOLTAIC ENERGY CONVERSION

AAA Solar Service and Supply
   http://www.rt66.com/aaasolar/

Alternative Energy Engineering, Inc.
   http://www.alt-energy.com/aee

American Solar Energy Society
   http://www.csn.net/solar/

Centre for Photovoltaic Devices and Systems
   http://www.vast.unsw.edu.au/pv.html

El Paso Solar Energy Association Home Page
http://www.realtime.net/~gnudd/react/epsea.htm

International Solar Energy Society
http://www.ises.org/

Mr. Solar's Home Page
http://www.netins.net/showcase/solarcatalog/

Maine Solar House
http://solstice.crest.org/renewables/wlord/index.html

National Renewable Energy Laboratory
http://www.nrel.gov/

N.C. Solar Center
http://www.ncsc.ncsu.edu

NW - SolarCell
http://www.seanet.com/Users/miknel/NW-SolarCell.html

Photovoltaic Technology
http://www.sandia.gov/Renewable_Energy/photovoltaic/pv.html

Solar Electric House, The
http://www.ultranet.com/~sda/bookinfo.html

Solar Energy
http://www.energy.ca.gov/energy/earthtext/solar.html

Solar Energy - Raymond J. Bahm
http://www.rt66.com/rbahm/

Solar Energy Network, The
http://204.214.164.14/indexa.html

Solarex of Frederick, MD
http://www.solarex.com

Solstice (Solar in Indonesia)
http://solstice.crest.org/renewables/indonesia/index.html

Sunny Gleason and Solar Powered Systems
http://www.cove.com/~sunny/

Sunrayce 95
http://www.nrel.gov/sunrayce/

World Solar Challenge
http://www.engin.umich.edu/solarcar/WSC.html

POWER

University of California Energy Institute
http://www.ucenergy.eecs.berkeley.edu/ucenergy

Sandia National Laboratories
http://www.sandia.gov/

U.S. Department of Energy (DOE)
http://www.doe.gov/

SOLAR THERMAL ENERGY CONVERSION

Sandia National Laboratory
http://www.sandia.gov/

SOLAR/OCEAN THERMAL ENERGY CONVERSION

Energy from the Ocean
http://zebu.uoregon.edu/ph162/l17small.html

Natural Energy Laboratory of Hawaii
http://bigisland.com/nelha/

THERMOELECTRIC

Thermoelectric Cooling (TEC)
http://www.trademart.com/aoc/tec.htm

Teledyne Brown Engineering - Hydrogen and Thermoelectric Generators
http://www.tbe.com/products/generators/generators.html

WIND ENERGY

American Wind Energy Association
http://www.econet.org/awea/
http://solstice.crest.org/renewables/awea/index.html

Bergey Windpower Co., Inc.
http://www.bergey.com

California Energy Commission
http://www.energy.ca.gov/energy/html/directory.html

International Wind Energy Associations
http://solstice.crest.org/renewables/wind-intl/index.html

Wind Energy in California
http://energy.ca.gov/energy/wind/wind-html/wind.html

Wind Energy Web Sites
http://www.strath.ac.uk/~cadx741/wind.html

World Power Technologies, Inc.
http://www.webpage.com/wpt

GLOSSARY

Alternator: An electromagnetic transducer used to convert mechanical energy, as from a water, wind, or steam turbine, to electrical energy in the form of alternating current and voltage.

Amorphous Silicon (a-Si): Silicon material manufactured at relatively low temperatures through a thin-film deposition process in which no crystalline structures are formed — a-Si has very low charge mobility and low efficiency compared to p-Si and x-Si material.

Armature: The moving portion of an electromagnetic transducer.

Array: As in *photovoltaic array*—a combination of modules and/or panels to make up an entire photovoltaic system producing a desired overall net power.

Battery: A series or parallel combination of voltaic cells—provides a higher voltage and/or current than a single cell.

Boule: An ingot or single-crystal silicon that has a cylindrical shape.

Breeder Reactor: A nuclear fission reactor that is cooled with a liquid metal (sodium) instead of water and operates at a higher temperature — breeds new fuel as it operates (Pu-239).

British Thermal Units (BTUs): One BTU is the amount of energy needed to raise the temperature of 1 pound of water $1°$ F.

Chemical Energy: Energy contained in the molecular structure of a substance — a form of potential energy which is converted to other forms of energy through some type of chemical reaction.

Chemical Energy Storage: As in *rechargeable batteries* — used to convert electrical energy into reactive chemicals. Battery chemistry is rebuilt (restored) during recharge thus storing energy in the chemistry that is able to produce electricity.

Chemical Vapor Deposition (CVD): Process in which silicon and dopants are introduced in gaseous form (silane, phosphine, etc.) into a special chamber where relatively low-temperature (250° C to 500° C) reactions take place that result in an even layer buildup of silicon on the substrate (glass, stainless steel, etc.).

Closed-Cycle System: An energy conversion system in which the source energy is transferred to an internal system where it is then converted to mechanical and electrical energy as in a closed-cycle OTEC system in which heat energy from sea water is transferred through a heat exchanger to a fluid such as alcohol, ammonia, or freon that is contained in a closed circuit and is used to drive a turbine.

Compressed-Air Energy Storage: Various sources of energy can be used to compress air in a large chamber thus storing energy — compressed air can be used to drive a turbine which drives a generator to produce electricity.

Condenser: That part of a heat exchanger or heat engine in which gas or vapor is cooled and converted to its liquid state.

Containment Vessel: A larger reinforced concrete structure in which the reactor vessel of a nuclear fission reactor is contained.

Control Rods: Used in nuclear fission reactors — made of materials that are able to trap and absorb neutrons. By maneuvering the control rods in or out of the core, the number of effective neutrons can be controlled and, therefore, the rate of chain reaction.

Covalent Bonding: Elements at the atomic level form into molecules using this process in which outer-shell electrons of each atom are shared by neighboring atoms.

Critical Reaction: In a nuclear fission reactor, the power level and a chain reaction being sustained at a constant rate.

Darrieus Wind Turbine: A type of vertical-axis wind turbine that has the appearance of a vertically stretched hoop — affectionately called an *eggbeater* — used to convert wind energy to mechanical and electrical energy.

Dopant: An element that is used as an impurity to make pure silicon into N-type or P-type material.

Electrochemical: Mechanism for creating electricity which involves the conversion of energy contained in chemical reactions to electricity.

Electrolysis: A process by which a DC electric current is passed through a conductive liquid (electrolyte) causing a chemical reaction to take place in which certain elements are removed from a compound or solution — electrolysis of water yields hydrogen and oxygen. Hydrogen is a fuel.

Electromagnetic: Mechanism for producing electrical energy which involves electromagnetic transducers called generators and alternators — magnetic field, conductors, and relative motion between the two are needed.

Electromagnetic Radiation: Electric and magnetic energy fields radiated from a source such as energy from the sun or a transmitting antenna.

Electrode: Terminal — conductors in an electrical system at which charges accumulate.

Energy: Comes from the Greek word *energeia*, which means activity — carries the connotation of doing work and is very simply defined as the capacity for doing work.

Evaporator: That part of a heat exchanger or heat engine in which gas or vapor is heated and converted from its liquid state to its gaseous state.

Field Coil: A coil that is used to create a magnetic field in an electromagnetic transducer.

Fission: The splitting of the atom's nucleus, and therefore the atom — used in nuclear power plants.

Flywheel Energy Storage: Often referred to as *inertial energy storage* since energy is stored in the inertia of a massive flywheel — heavy discs that absorb energy as they are caused to spin at tens of thousands of RPM. Higher speed = more stored energy.

Fuel Assembly: Made up of a large bundle of individual long tubular fuel pins, commonly called fuel rods (typically 236 fuel rods per fuel assembly) — a fuel rod will contain approximately 250 individual fuel pellets (each pellet about 1 cm in diameter and 1.5 cm long).

Fumarole: A venting of steam from the earth — a plume of steam.

Fusion: A process in which atoms are forced together to form heavier atoms — extremely high temperatures are required (100,000,000° C) — as in a nuclear fusion reactor — none in existence.

Generator: An electromagnetic transducer that converts mechanical energy to electrical — involves the use of a magnetic field, conductors, and motion.

Geopressurized Water Energy: Another geothermal source of energy that is being pursued by geologists the world over — water trapped in huge chambers or reservoirs deep in the earth's crust under high pressure and temperature.

Heat Engine: Any device that develops mechanical energy from thermal energy — efficiency of the heat engine is determined by the amount of thermal energy that is actually converted to mechanical energy as compared to the total amount of thermal energy applied to the engine. Examples: jet engines, diesel engines, Stirling engines, and steam engines.

Heliostat: A panel of reflectors that are used to direct the sun's energy at a central receiver unit where extremely high temperatures are produced.

Hot-Rock Geo-Energy: Closely related to hydrothermal energy in that extremely hot rock, deep in the earth's crust, is used to superheat water turning it into highly pressurized steam.

Hydro: Greek, referring to water or liquid.

Hydro Energy Storage: Stockpiling a liquid, usually water, in a holding area at a higher elevation than its use.

Hydropower: Power produced or contained in the kinetic energy of water or other liquids as in hydroelectric power.

Hydrothermal: Heat energy contained in hot liquids or expanding gas such as hot water and steam — as in geothermal, energy is manifested in two general ways: (1) as hot springs and geysers, and (2) as dry steam.

Impulse Turbine: Operates as high-speed water, via a jet nozzle, is sprayed at vanes or cups on the rim of a wheel.

Insolation Level: The amount of solar power impinging on a square meter of earth's surface — expressed in watts/meter.

Inverter: An electronic circuit that is able to convert DC to AC.

Joule (J): One watt-second (1J = 1W x 1s).

Kinetic Energy: Comes from the Greek word *kinetikos* which means putting in motion — the energy contained in a moving mass — work is being done.

Light Water Reactor (LWR): Nuclear reactor which uses pure water as the main coolant for the reactor itself — two main types: the boiling water reactors (BWRs) and the pressurized water reactors (PWRs).

Module: As in *photovoltaic module* — a combination of individual PV cells that produces a desired voltage and/or current. Modules are then used to make panels.

N-Type Semiconductor: Semiconductor material that has been doped with a pentavalent impurity (5 electrons in outer shell) to create a surplus of free electrons in the material. Eight electrons are needed to provide a full bond but in this case there will be nine — four from silicon and five from the impurity which releases one electron as free.

Nuclear Energy: The potential energy contained in the nucleus of an atom — the binding energy of the neutrons and protons.

Ocean Thermal Energy Conversion (OTEC): Using the solar energy stored in the ocean to operate a low-temperature heat engine (turbine) to produce electricity.

Open-Cycle System: An energy conversion system in which the energy is converted directly from the source as in an open-cycle OTEC system in which the warm sea water is turned to steam to drive a turbine.

Oscillator: An electronic circuit that creates an alternating current and voltage from a DC source — used in computers, radio receivers, transmitters, cell phone, etc.

Panels: As in photovoltaic panels — mounted grouping of modules to produce a desired voltage and/or current. Panels are used to make arrays.

Peltier Effect: The cooling and heating effect of a powered thermocouple — discovered in 1834 by James C. Peltier.

Photoelectric: Mechanism for producing electricity in which energy from the sun in the form of photon bombardment provides energy to electrons in semiconductor materials and forces them to flow.

Photovoltaic (PV) Cell: An electrical transducer unit that produces a difference of potential as a result of photon bombardment — each cell typically produces around 0.5 V in bright direct sun.

Piezoelectric: Mechanism in which pressure, torque, or vibration is applied to certain crystalline minerals causing electricity to be produced. (Also known as the *piezoelectric effect*.) Discovered by two French scientists, Pierre and Jacques Curie, in 1880.

Polycrystalline-Silicon (p-Si): Silicon material that is manufactured at a lower temperature than single-crystal silicon in such a way as to produce many crystalline structures throughout the material — p-Si has less charge mobility and is less efficient than x-Si material.

Potential Energy: Relational energy — held by a stationary mass that is being acted upon by a force in relationship to a certain point. Has the potential for doing work.

P-Type Semiconductor: Semiconductor material that has been doped with a trivalent impurity (three electrons in outer shell) to create a hole in the covalent bonds between the atoms. Eight electrons are needed to provide a full bond but in this case there will only be seven — four from silicon and three from the impurity.

Reaction Turbine: Has propeller-like rotors that are installed within a pipe or cylinder — water is forced through the blades causing them to turn.

Reactor Vessel: The core of the nuclear fission reactor is contained in this heavy stainless steel chamber.

Reciprocity: The property of many transducers whereby the transducer is able to function bidirectionally in converting one form of energy to another — i.e. a loudspeaker can be used as a microphone and a DC motor can be used as a DC generator.

Renewable Energy: a.k.a *Sustainable Energy*, or energy sources that are not depleted but are renewable by nature.

Seebeck Effect: The thermoelectric mechanism for producing electricity named in honor of Thomas Johann Seebeck who discovered it in 1821.

Single-Crystal Silicon (x-Si): Silicon material that is grown in a high-temperature oven forming a cylinder-shaped ingot called a boule.

Solar Cell: See *Photovoltaic Cell.*

Solar Thermal Conversion (STC): Employs the use of concentrating solar collectors that focus and convert the sun's energy to very high temperatures.

Sonic Transducer: A device that is capable of converting audible sound energy (sonic) into electrical energy and/or vice versa.

Stand-Alone (SA) System: An electrical energy producing system installed with all of the wiring, associated electronics, and means for energy storage —independent of a utility company.

Stator, Stator Winding: A stationary or fixed coil in an electromagnetic device.

Subcritical Reaction: In a nuclear fission reactor the power level and chain reaction is decreasing.

Sunrayce: A biennial event sponsored by the U.S. Department of Energy (DOE) and managed by one of its national laboratories — co-sponsored by many corporations such as General Motors, who began the event, Electronic Data Systems Corporation, Midwest Research Institute, the Environmental Protection Agency, the Society of Automotive Engineers, Sandia National Laboratories, and the National Renewable Energy Laboratory.

Supercritical Reaction: In a nuclear fission reactor the power level and chain reaction are increasing.

Synchronous Inverter: A DC to AC converter that is tied to a utility line and can feed surplus energy back into the utility system — used in Utility-Interactive (UI) systems.

Thermal Energy: Energy produced as a result of friction, chemical reaction, or nuclear reaction, and is characterized by a change in temperature.

Thermocouple: An electrical transducer that converts a difference in temperature to a difference in potential (voltage). See *Thermoelectric*.

Thermoelectric: Mechanism in which two conductors, made of different metals, are joined on each end to form a loop and the temperature of one end is much different than the other which produces a small electrical current.

Thermoelectric Generator: A system in which a heat source is applied to a thermopile to produce direct current (DC) electricity.

Thermonuclear Reaction: Nuclear fusion reaction involving the collision of atoms — with each collision, and subsequent fusion, tremendous amounts of energy in the form of heat are released. As the chain reaction continues, the very high temperature is sustained by the reaction itself.

Thermopile: Thermocouples connected in series for more voltage and in parallel for more current.

Transducer: Device that transforms energy from one form to another.

Ultrasonic Transducer: A device that converts inaudible high frequency sound, or mechanical vibration, to an electrical voltage, or signal and/or vice versa.

Utility-Interactive (UI) Systems: Electrical energy producing systems that are tied into the utility lines — uses the utility lines for surplus energy storage. No batteries are needed for energy storage.

Valence Shell: The outer shell of atoms — *valence number* is the number of electrons in the outer shell.

Voltaic Cell: A single electrochemical system that produces a specific voltage based on the chemistry involved — the smallest functioning unit that produces a certain potential in an electricity-producing system.

Watt (W): The unit of electrical power — a rate at which electrical energy is expended. 1 W = 1 joule of energy expended every second = 1J/1s.

Watthour (Wh): A measure of consumed energy — power times hours of use — i.e. 1 kWh = 1000W x 1h, meaning 1 kilowatt of electrical power was used for 1 hour. The same as 200W being used for 5 hours.

INDEX

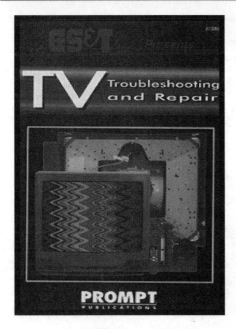
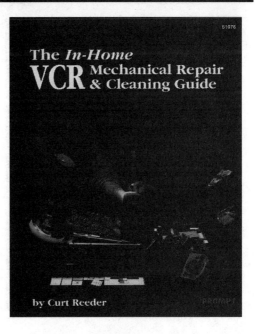

PROMPT® Publications

Speakers for Your Home and Automobile

How to Build a Quality
Audio System

*By Gordon McComb, Alvis J. Evans,
and Eric J. Evans*

The cleanest CD sound, the quietest turntable, and the clearest FM signal are useless without a fine speaker system. With easy-to-understand instructions and illustrated examples, this book shows how to construct quality speaker systems.

$14.95

Paper/164 pp./6 x 9"/Illustrated
ISBN#: 0-7906-1025-6
Pub. Date 10/92

Sound Systems for Your Automobile

The How-To Guide for
Audio System
Selection and Installation

By Alvis J. Evans and Eric J. Evans

Whether you're starting from scratch or upgrading, this book will show you how to plan your car stereo system, choose components and speakers, and install and interconnect them to achieve the best sound quality possible.

$16.95

Paper/124 pp./6 x 9"/Illustrated
ISBN#: 0-7906-1046-9
Pub. Date 1/94

Call us today for the name of your nearest distributor.

800-428-7267

PROMPT® Publications

"Is This Thing On?"
Sound Systems for Your Business,
School, and Auditorium
By Gordon McComb
Is This Thing On? takes readers through
each step of selecting components, install-
ing, adjusting, and maintaining a sound sys-
tem for small meeting rooms, churches, lec-
ture halls, public-address systems for
schools or offices, or any other large room.
$14.95
Paper/6 x 9"/128 pp./Illustrated
ISBN #0-7906-1081-7
Pub. Date 6/96

TV Video Systems for the Hobbyist & Technician
The How-To Guide to Installation,
Operation & Troubleshooting
By L.W. Pena and Brent A. Pena
TV Video Systems explains the different
video programming systems, how they are
installed, their advantages and disadvantages, and how to detect problems and add
accessories. Written to inform the reader
about the choices available to receive TV
signals.
$14.95
Paper/6 x 9"/128 pp./Illustrated
ISBN #0-7906-1082-5
Pub. Date 5/96

Call us today for the name of your nearest distributor.

800-428-7267

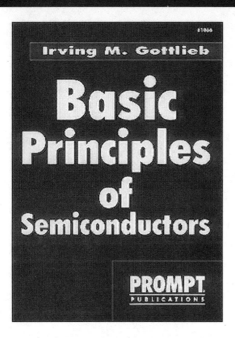
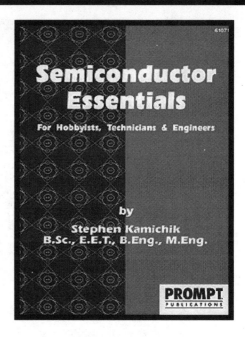

PROMPT® Publications

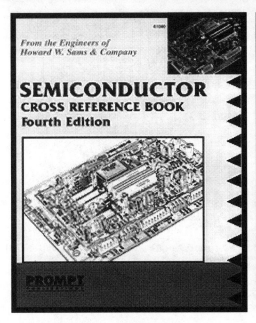

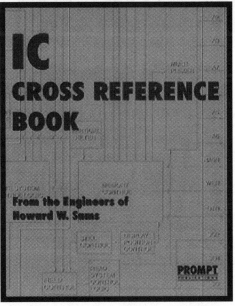

Semiconductor Cross Reference Book
Fourth Edition
By Howard W. Sams & Company
From the makers of PHOTOFACT® service documentation, the *Semiconductor Cross Reference Book* is the most comprehensive guide to semiconductor replacement data. The volume contains over 490,000 part numbers, type numbers, and other identifying numbers.
$24.95
Paper/678 pp./8-1/2 x 11"
ISBN#: 0-7906-1080-9
Pub. Date 8/96

IC Cross Reference Book
By Howard W. Sams & Company
The engineering staff of Howard W. Sams & Company has assembled the *IC Cross Reference Book* to help you find replacements or substitutions for more than 35,000 ICs or modules. It has been compiled from manufacturers' data and from the analysis of consumer electronics devices for PHOTOFACT® service data, which has been relied upon since 1946 by service technicians worldwide.
$19.95
Paper/168 pp./8-1/2 x 11"
ISBN#: 0-7906-1049-3
Pub. Date 5/94

Call us today for the name of your nearest distributor.

800-428-7267

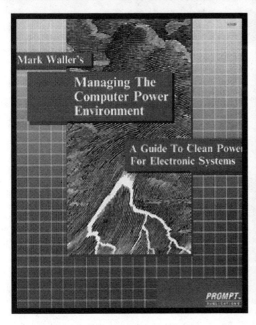

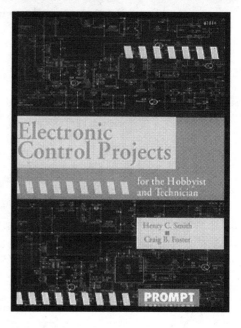

PROMPT® Publications

Power Supplies
Projects for the Hobbyist
and Technician
By David Lines

Power supplies, the sources of energy for all electronic equipment, are basic consideration in all electronic design and construction. This book guides you from the fundamentals of power supply components and their functions to the design and construction of a power supply system.
$10.95
Paper/6 x 9"/96 pp./Illustrated
ISBN #: 0-7906-1024-8
Pub. Date 12/92

Simplifying
Power Supply
Technology
By Rajesh J. Shah

Simplifying Power Supply Technology is an entry point into the field of power supplies. It simplifies the concepts of power supply technology and gives you the background and knowledge to confidently enter the power supply field
$16.95
Paper/6 x 9"/160 pp./Illustrated
ISBN #: 0-7906-1062-0
Pub. Date 3/95

Call us today for the name of your nearest distributor.

800-428-7267

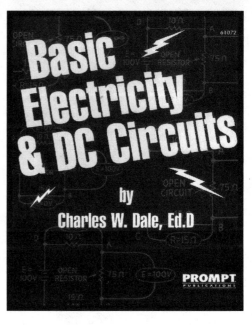

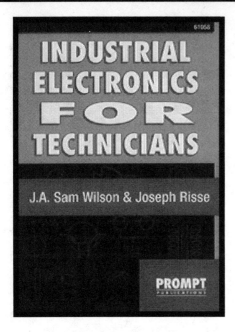

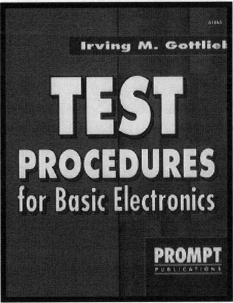

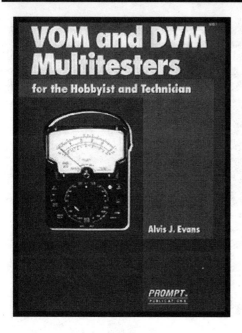

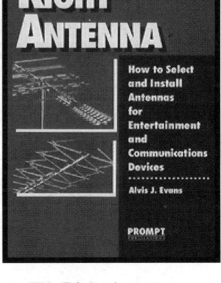